Research for Development

The series Research for Development serves as a vehicle for the presentation and dissemination of complex research and multidisciplinary projects. The published work is dedicated to fostering a high degree of innovation and to the sophisticated demonstration of new techniques or methods.

The aim of the Research for Development series is to promote well-balanced sustainable growth. This might take the form of measurable social and economic outcomes, in addition to environmental benefits, or improved efficiency in the use of resources; it might also involve an original mix of intervention schemes.

Research for Development focuses on the following topics and disciplines:

Urban regeneration and infrastructure, Info-mobility, transport, and logistics, Environment and the land, Cultural heritage and landscape, Energy, Innovation in processes and technologies, Applications of chemistry, materials, and nanotechnologies, Material science and biotechnology solutions, Physics results and related applications and aerospace, Ongoing training and continuing education.

Fondazione Politecnico di Milano collaborates as a special co-partner in this series by suggesting themes and evaluating proposals for new volumes. Research for Development addresses researchers, advanced graduate students, and policy and decision-makers around the world in government, industry, and civil society.

THE SERIES IS INDEXED IN SCOPUS

More information about this series at http://www.springer.com/series/13084

Annalisa Dominoni

Design of Supporting Systems for Life in Outer Space

A Design Perspective on Space Missions Near Earth and Beyond

Springer

Annalisa Dominoni
Design Department
Politecnico di Milano
Milan, Italy

ISSN 2198-7300 ISSN 2198-7319 (electronic)
Research for Development
ISBN 978-3-030-60941-2 ISBN 978-3-030-60942-9 (eBook)
https://doi.org/10.1007/978-3-030-60942-9

This Springer imprint is published by the registered company Springer Nature Switzerland AG
The registered company address is: Gewerbestrasse 11, 6330 Cham, Switzerland

To my Beloved

Preface

In this book, the newly emerging discipline of *Space Design* is presented with the aim to enhance the research of behaviour and well-being for space travellers, both astronauts and tourists. Space scenarios and technologies will increase in the next future and have a strong impact on how people behave and perform. Also, we can look at the International Space Station (ISS) as the best example of *living green* in a *sustainable way* and in a community of international cooperation without cultural boundaries.

The author, thanks to her 20-year experience in researching and designing habitation modules and tools that have been used by astronauts on board the ISS, has helped assert the Design's role and purpose in the Space field. She also created and is currently leading the first and unique International Master of Science Course in *Space Design* in the world, *Space4InspirAction (S4I),* recognized and supported by the European Space Agency (ESA), in which research and teaching are strongly connected with the industry with the aim to generate new disruptive ideas for Space.

The *Space Design* acknowledgment is due to the evolution of the strategic programmes for interplanetary exploration in which human beings will spend increasingly longer periods of time in reduced gravity and extreme conditions and need a higher level of comfort. It is certain today—with the experience collected in few tens of years on board the International Space Station (ISS)—that by increasing the well-being of the crew, the functionality of the activities and the usability of interfaces and tools for astronauts, we can achieve better goals that directly affect the success of a space mission. We have to consider also that already now, and in the future, not only super-skilled astronauts with a strong training behind, but also researchers, scientists and, not least, tourist, will have the chance to experiment space travels. This means that the whole environments must be totally re-thought and re-designed according to the physiological and psychological needs of the space travellers, as well as objects, facilities and tools for the daily life activities—living, working, sleeping, eating, taking care of ourselves, resting, etc.—which should make the life in Space easier, trying to reduce the difference between *living in Space* and *living on Earth*. For these reasons, the role of Design in Space is important.

The book gives the opportunity to know more about the importance of Design researches and solutions in Outer Space for the support and development of human space missions. International experiences and projects of Space Design, researches and on-orbit experiments with astronauts on the International Space Station (ISS) led by the author will be presented through case studies to clarify and emphasize the peculiarity of Design skills in Space, considering specific requirements and constraints to face with. The challenge is to transform the environmental extreme conditions from a *limit* to an *opportunity*—as taking advantage from the reduced gravity—to improve the quality of life in Space with emotional language and usability tools.

In the first chapter *The Strategic Role of Design for Space,* the focus is on the core of the discipline *Space Design* which is able to speak both the different languages of science and beauty, find disruptive solutions that could become spin-offs and spin-ins of technologies and behaviours from Space to Earth and vice versa, putting at the same time the human being in the centre. What Design does for Space? What are the Design domain fields for Space? How *Space Design* can be considered a driver for the New Space Economy? Which are the specific methodologies for *Space Design*?

In the second chapter *Living in Space by the Lens of Design,* all the activities carried out by the astronauts on board the ISS are considered according to the peculiarities of the environment, as the transition to microgravity and the confinement, looking at the countermeasures adopted to limit the negative effect living in Space from a Design point of view.

In the third chapter *Research and Design for Space Life,* the birth of the new discipline in *Space Design* is described through international experiences of researches and projects developed by precursors, architects and designers and by the author from the beginning of the *Space Era* to now and beyond.

In the fourth and last chapter, five case studies are presented to underline the strong connection between research, project development and academic activities led by the author designing for Space. From experiments of clothing support systems and tools for human space missions on board the ISS to the design for new space stations, the conquest of Space is now a reality that is taking shape in an increasingly concrete and fascinating for the younger generations, and projects such as *VEST*, *Couture in Orbit*, *Space4InspirAction (S4I)*, *Fashion in Orbit*, *Space Fashion Design* and *Moony* for *Igluna*, described in this book as case studies, help to direct young people and the whole society towards a shared and livable interplanetary future and in which the Design will always have a fundamental role, if not indispensable.

The goal of the book is to demonstrate the important role and diffusion of Design in Space and to allow readers to enter into new scenarios imagining how we could live in the near future. Design contribution and skills can generate innovation, thanks to a strong capacity of visioning, taking inspiration by Space technology to find spin-offs in other fields, and also spin-ins looking at Earth solutions.

The book is foreseen for researchers and students but could be interesting also for a broad readership attracted by Space, innovation and future scenarios, mainly the young generation.

Just a little note: readers will find that in the book the words *Space* and *Design* are written with capitalised terms when decontextualized could be misunderstood, the first with environments and places, instead of cosmic Space, the second with project, instead of discipline and method inherent the Design.

Milan, Italy Annalisa Dominoni

Contents

Chapter 1
The Strategic Role of Design for Space

1.1 The Fascination for Space

On 20 July 1969, the whole world was facing the Moon. It seemed impossible that the Apollo 11 could bring human beings to our satellite and bring them back safely. The landing happened in the evening, I remember it because I was at the sea in Liguria, the sky was clear, and I was very curious to see the astronauts touch the lunar ground. To the *naked eye*. That's right. I was 4 years old, and I was sure that going out in the garden and sitting comfortably in my small chair I would see them, Armstrong and Aldrin, walking and running on the surface of white iridescent dust. And there they were, I only saw shadows, but I was sure I saw them. I was very excited and very happy. Of a deep happiness that I still remember. This episode was the beginning of my attraction to the cosmos and the stars.

A long time later I realized that Space that summer had not only fascinated me, along with other children, but had contributed to the planetary explosion of the *Space Age*, a period of contemporary history born between the 1950s and 1960s parallel to the human space exploration that influenced the common imagination and directed the technological development of humanity. All the creative currents, from literature to music, from architecture to cinema and from fashion to art and design, have undergone in those years the charm of Space producing powerful icons: just think of the film *2001 A Space Odyssey*, by Stanley Kubrick (Fig. 1.1), the comic book *Barbarella* by Jean-Claude Foret, the fashion of the *Sputnik Girls* by Pierre Cardin or the design of the television *Brionvega Doney 14* by Marco Zanuso (Fig. 1.2), which looks like a helmet for astronauts. Objects with lunar shapes that interact with environments and architectures inspired by Space's iconography, and in which characters dressed in white and silver seem to move raised from the ground, were the tangible expression and widespread optimism for the conquest of Space that brought with it great expectations of growth and development.

A. Dominoni, *Design of Supporting Systems for Life in Outer Space*, Research for Development, https://doi.org/10.1007/978-3-030-60942-9_1

Fig. 1.1 *The Fascination for Space* is expressed in many creative fields as movies that have predicted some aspects of space travels: here an interior view of *2001: A Space Odyssey*. Credits by Warner Bros

The *Space Age* also coincides with the economic boom, with the birth of the *Pop Art*, which celebrated irrepressible consumerism made of products dressed in attractive and redundant communication, and with the triumph of plastic over other materials, for practicality, low cost and the ease with which it allows to generate smooth, curved and aseptic surfaces like those of a spaceship: round shapes, soft and organic, replaced rigid and squared volumes of the traditional style, thanks to technological innovation (Dominoni and Quaquaro 2017a; b). From this brief excursus, it is evident that initially the Space inspiration, as far as Design is concerned, had its most powerful expression in the definition of a *style innovation*, through a language that used mainly shapes and colours that reflected the imaginary of the *Space Age*. Then, it is a *technological innovation*, the plastic, which determines the possibility of creating new soft and organic shapes. The style intersects with the technology and breaks the pre-constituted order: it is not necessarily the form that is decisive in the design process, even technology can suggest new forms that were not even possible to imagine in advance. Thanks to the research of designers who, in close contact with the industry, first of all in Italy the Kartell, begin to experiment with material and molds going to the limit and beyond the feasibility, the Design explores new expressive languages and gives form to new artifacts that determine a real revolution: new types of objects are added to existing ones, and many objects built with other materials are now made of plastic to the advantage of a significant reduction in costs and a greater lightness and flexibility of use. This innovation determined a wide spread of objects and democratization of Design: they increased the performance and aesthetic qualities of objects and we saw a change in lifestyles. But the most important point related to the plastic is that

Fig. 1.2 Icons of style, as the *Brionvega Doney 14* of Marco Zanuso and the *Sputnik Girls* with vinyl eyewear of Pierre Cardin, have been inspired by *Space Age* introducing a new design and fashion language. Collage by the author

it was a *revolutionary innovation* pushed from a vision, and Design is a central element in every typology of innovation, but overall in this one (Dominoni 2009) that make the difference. Because Design approach is able to add a meaning value to innovations translating them in products and services which are bearers of dreams, images, expectations of more well-being. The results of the revolutionary innovation are breakthrough and *disruptive ideas* which create a scenario completely new, also modifying the existent consumers' behaviour on Earth, and why not, in Space.

1.2 Space Design for Disruptive Ideas

This book allows me the opportunity to present the newly emerging discipline of Space Design, aimed both to enhance the research of wellness in Outer Space, increasing the comfort of the space travellers, and to generate innovation through the contamination of living in Space and living on Earth, crossing different knowledge and fields of applications, as behaviours and technologies that can be taken as inspiration by Space and translated to Earth, and vice versa, becoming spin-offs and spin-ins.

In this first chapter, I explain through experiences, projects and researches *The Strategic Role of Design in Space* introducing aspects, methodologies, principles and tools, specific of the discipline, that will be treated with more comprehensively in the following chapters. My aim is to give an idea more precise of what Design is, what it does on Earth and what it can do for Space, and in what domain fields it can intervene, in which contexts its competences and skills are required.

Design is a discipline and a profession with a great capacity to answer to the needs of the human being, as well as those of the industry, and with the power to influence and change the behaviours of the society. As architect and designer, I believe we should design more *conscious* objects and environments. The objects which have more success are those able to arouse emotions and make people think, when they become a source of inspiration for people. The environments which obtain more attention are those where people are at ease and where visual and sensorial stimuli produce a sense of well-being and happiness that is imprinted in the memory through the shapes and light that express the architecture of the space.

The Design domain fields for Space appears to be very huge and variegated embracing the world of artifacts and the human beings. Space is a very stimulating world to explore for a designer who wants to deal with the new, the unknown, and who is able to imagine *disruptive* environments and objects, which do not belong to our daily life experience on Earth, but which could be very suitable and useful, and also *beautiful*, for a new generation of people who will inhabit and perform them.

Having to first deal with survival and extremely technical issues, the space industry is dominated by scientists, coming mainly from medicine and engineering, who seem do not understand the importance of the Design in increasing the well-being of the crew and, as consequence, the whole mission success.

In *Industrial Design for Space* (Dominoni 2002), I introduced the assumptions to create a dialogue between Industrial Design and Space Industry to demonstrate that Design could have a fundamental and strategic role for Space in designing new environments and tools that can facilitate human movements and activities, foreseeing new ways of using instruments and new gestures in relation to the extraordinary conditions, and choosing scientific and industrial partners with the specific know-how required for each project. Approaching and developing the first researches, I realized that there were necessary new tools to face a context so different and challenging as Space is, and I started creating and experimenting a new methodology specific for Space Design that I called *Use and Gesture Design (UGD)*. This method is based on the interactions between the human being and the objects that design a new way to decode gestures, movements and postures in microgravity and in confined environments.

My intention was to convey the importance of observing people and reality, alone and in the environment in which they live, move and interact with each other and things, to design *conscious objects and environments* that can correspond to the character, desires and needs of each of us, on Earth and in Space. The designer is an acute *interpreter* who registers the world as it is, but equally capable of developing a new thought. Thanks to its ability to observe the reality that surrounds us and know-how to interpret the many signs that every day we receive from the environment, watching how people behave, move and interact with each other and interpret in absolutely unique way life experience.

Through the lens of Design, we can translate latent needs, not yet expressed by people in general, and astronauts, in particular, through their behaviours, spontaneous gestures, and also, *improper use of objects*, into projects to make people feel good, in *ethical products* that really could help astronauts to live and work better, and which have more than one kind of utility, which it is emotional in nature, as well as functional. We must not underestimate the decisive role which has emotions in the choices that we do every day and that lead us to surround ourselves of objects charged with sense able to arouse in us pleasant memories of past experiences or reactivating atavistic memories that make us act on impulse, with the belly and not with the brain. Therefore, the skills of a designer have to include a mix of intuition and psychology, to investigate the unconscious and the most hidden parts of the human being learning how to use thinner languages that make visible those parts that are expressed through gestures. This practice brings out an increasing attention to those factors related prevalently to the emotion and psycho-perceptive sphere combined with an artistic approach based on *aesthetic research* that places the human being at the centre and makes use of expressive languages aimed to enhancing the *poetics* of the objects and the environments.

Disruptive ideas arise from the capacity of Space Design to develop an attitude to the project looking beyond the functional aspects of the products, or the innovation technologies, focusing on how products and environments can reflect lifestyles, values and aspirations.

Following the evolution of the designer figure, both researcher and professional, appears evident a constant that concerns the multidisciplinary approach to the projects, in which the formal and functional qualities of the products interact transversally with coming knowledge by engineering, human sciences, technology or economy innovation, bearing in mind the specificity that has always distinguished Design, that means the attention to people, the interest in the human factors as well as irrational and sensorial perceptions, the mental mechanisms of childhood that with time hide in the unconscious, the influence of experience on our view of the world. Considering the complexity of managing projects for Space, in another part of the same book (Dominoni 2002), I suggested the strategic importance to create cross-disciplinary teams of specialists with different competencies, together with industries with different know-how, of both Space and private sectors, able to face with different typologies of products and tools and find innovative solutions.

For the designer, accustomed to dealing with productive "terrestrial" realities, the Design Research in Space is possible only with the constant presence and participation of industries, institutions and R&D centres able to provide all the necessary information and know-how to face from time to time the design assignments. As a consequence, the group is a natural amplifier and accelerator of the creative process which serves as a *collective brain*, with much higher potential of the single individual brain. Therefore, the advantage resides in the characteristics of speed and richness of response to the demand for innovation (Dominoni 2009).

Cross-collaboration and interdisciplinarity are becoming more and more prominent in the last years in both academic and industrial conversations. As within western societies knowledge has been subject to fragmentation, cross-disciplinarity is considered as an attempt to overcome fragmented thinking and develop holistic modes of enquiry, decision-making and practice (Kline 1995). Indeed, it was broadly discussed the need to cross-disciplinary boundaries in order to tackle contemporary complex problems, such as climate change (UNESCO 2015).

Having analysed in depth the group dynamics in the past, I can affirm that the context of cross-disciplinary innovation teams is particularly interesting as a slightly more open door for designers to step in the space industry. In recent years, there has been a growing evolution in the Space field, triggered by future planned human space missions, which once again propose the desire for exploration as the main engine, now driven principally by the private sector which looks at space tourism as a propellant to increase interest and budget. With a permanent settlement on the Moon and a manned landing on Mars as goals to support pioneer ambitions and efforts, a great boost is given to the research, in order to fill the gaps that still make these perspectives impracticable. Following a trend already widespread in several industrial sectors, space agencies are turning to the adoption of cross-disciplinary teams as a tool to stimulate innovation.

Design is, indeed, synonymous of innovation focused on the human needs, and by time it developed including knowledge from engineering, innovation economics and technology, in order to better interface with science as the basis for novel applications (Dominoni 2009).

Design for Space means redefining the task of industrial design in this particular context, find interlocutors and possible areas of convergence, methodological and operational tools in order to offer its expertise to space agencies and industries and contribute to solving the problems of living and working in confined environment and in microgravity conditions with innovative projects.

The design activity, even if indispensable, is not sufficient for the success of the project if it is not integrated with the participation of these other actors. Once this happens, small agile and cross-disciplinary teams are nowadays recognized as the main drivers of innovation.

From the beginning, I was convinced that the role of the designer was to build a *bridge* between different disciplines and knowledge to develop new products and tools to allow astronauts to live and work in better conditions in Space, and specifically on board the International Space Station (ISS). The task of Design, in this particular context, is to increase the comfort and well-being of the crew designing environments, equipment and tools in order to facilitate human movements and improve the performance of the various activities, in favour of greater productivity and quality of life on board. There was to build a whole new world.

1.2.1 The Design Domain Fields for Space

Design can intervene in the design process for Space at various levels and scales: designing beautiful and functional environments to make astronauts feel better when living and working on board the ISS, maybe changing the internal configuration in order to increase the habitability performances, or introducing innovative lighting system to change the perception of space and balance the circadian rhythms that in Space are altered for the lack of natural stimuli; proposing equipment to facilitate the various activities, for example, fitness in microgravity, in order to maintain bone and muscle mass, but also try to make physical exercises more attractive introducing entertainment; creating specific tools for the preparation, consumption and conservation of food, or objects to facilitate self-care and guarantee a good level of personal hygiene; increasing the performance of the working instruments to support astronauts during the various tasks; and improving the leisure time favouring both the needs of privacy and conviviality or imagining new way to make easy the communication with family and friends. These are only a few examples of the activities, needs and fields of applications the Space Design face with that in the absence of gravity are much more complex and require the designer greater attention to the human body—from the physiological, perceptive and psychological point of views—which responds in a different way to the various particular extreme conditions, and to the interaction with objects and environment.

There is no substantial difference between the project of a designer who creates new objects and structures for Space compared to those who are dedicated to making a better place the Earth. In my opinion it worth, in both the cases, the historic slogan of Ernesto Nathan Rogers *from the spoon to the city* in which the Milanese architect and theoretician of architecture intended to express the breadth of the territory project of our profession, which allows to act simultaneously on different scales of complexity in relation to the human being. Of course, in Space the movements and gestures that each of us acquires living and having direct experience through the environment surrounding are altered, as well as the shape, the posture and physiology of the body, which in microgravity undergo obvious transformations. Then our *spoon* will have surprises for us, in case we want to try to use it to consume a breakfast of milk and cornflakes, because we would see small white spheres together with flakes of cereals start to float and go around the International Space Station (ISS), most likely clogging the air filters. Maybe a straw is better. Or maybe, even better would be that the designer would think how to solve this need through a new thought that does not refer to the experience of all the days on Earth, but based on extraordinary conditions in Space, and situations sometimes much closer to dreams than to reality.

When we dream, we can do somersaults in the air just like the first astronauts experimented during the first Skylab human space mission, or we can have fun breaking an egg and watching the egg white that remains around the yolk as if it were the most natural thing in the world (Dominoni 2015). Design can improve the *function* and the interface *usability* of the electrical appliances and the control and navigation systems as well as the *quality* of the lighting systems. For example, how we could wash garments in Space? When I led a feasibility study for a *space washing machine* to be used on board the ISS, I asked myself if it was possible to think of washing clothes without water and using the vacuum outside the spaceship. I have chosen as industrial partner of the project Electrolux Zanussi, a big international company in electrical appliances, to give me the possibility to face new technologies and imagine new ways to clean garments in orbit. I built a cross-disciplinary team to have the terrestrial know-how related to the washing

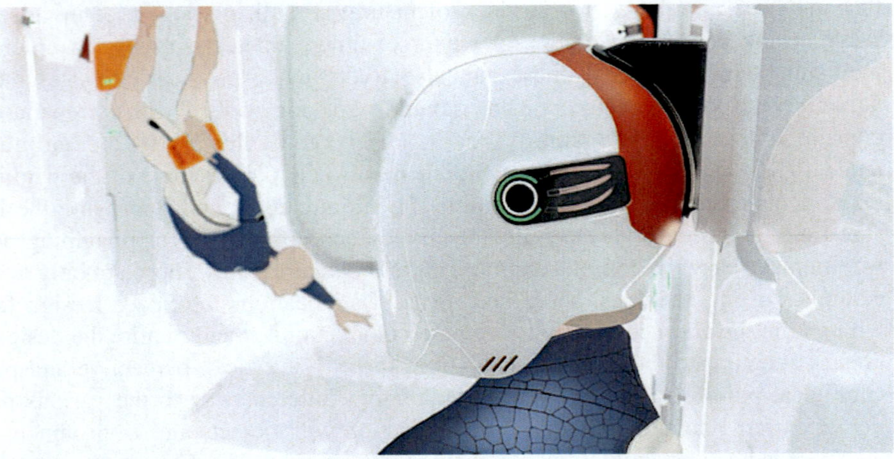

Fig. 1.3 The well-being of the astronauts depends also on natural stimuli that in confined environments should be recreated as using lighting effects to balance circadian rhythms. Credits by *Space4InspirAction (S4I)* Course, 2nd edition 2018, Politecnico di Milano

technologies together with chemists, to imagine new enzymes and molecules to face the vacuum alternatives, and space engineers, to guarantee the *spatialization* of the new washing machine for Space.

The study of *lighting*, as it happens already for the design of both small rooms and big office spaces that not always can rely on the outside light, must take into account to balance the circadian rhythms. Natural light is fundamental to regulate sleep and waking, and its lack in confined environment causes strong alterations of the physiology of the human being generating insomnia and general impatience. A *good design* has to develop illuminating bodies with variants, in terms of temperature of light and luminous intensity, similar to natural ones (Fig. 1.3). Light can also create space partitions defining private rooms or can change the perception of the environments, overall if in isolation conditions. Using lighting effects, we can recreate the Sun that filters through the windows reflecting multiple and vibrant shadows. Also in the science fiction imagination, there are solutions that seek to recreate environmental stimuli natural, as you can see in the movie *Solaris* by the Russian director Andrei Tarkovsky, in which the sound of the wind was reproduced through strips of paper placed under the aeration fans of the space station. Coming back to the reality, in Space the crew members often in the absence of adequate solutions try to create a better environment through the most unthinkable devices that they have available as the cosmonaut Anatoly Berezovoj who in 1982 spent part of his 7 months on board of Salyut 7 listening to a cassette player in which they had been recorded the sounds of woods, birds, wind and of the storming of the leaves, in short, the terrestrial sounds (Clark 1988).

The *well-being* of astronauts also depends on a number of natural environmental stimulations—such as the light of Sun, colours, air, wind, temperature variations,

Fig. 1.4 The role of Design for Space is to increase the comfort of the crew imaging new products and habitation modules according to the extreme environmental conditions. Credits by *Space4InspirAction (S4I)* Course, 1st edition 2017, Politecnico di Milano

scents, flavours and sounds—that normally we live and feel on Earth and that are able to stimulate the vital functions, and that unfortunately in Space the artificial environment of the habitation modules is not still able to reproduce, and over time provokes monotony, a loss of widespread interest and also depression. The human being, under normal conditions, reacts to these stresses tuning its balance with nature and the surrounding environment. For these reasons, the lack of a single natural stimulus can therefore compromise biological and mental health.

Increasing the scale of the project, the Design is fundamental to design means of transport to Space, as the fleets of the launchers Apollo, Shuttle and the new generation Falcon and Crew Dragon, but also new space stations (Fig. 1.4), Moon and Mars bases, and rovers to explore the extreme territories. In these contexts, the designer contributions can regard the shape of the external structure, the interior habitat for the crew and the furniture defining all the details, such as the coatings, the colours, the materials, the structures of the seats that will welcome the astronauts on the space vehicles, or the restraint systems to anchor inside the ISS, and even clothing for living and working in confined environment and in microgravity conditions (Souza et al. 1991–1998). At this scale, the most important aspect that affects the design choices is precisely the confinement and the lack of gravity (Stuster 1986). The designer deals with very small spaces that need to be used intelligently to allow crew to live and work in a *comfortable* and *efficient* way, both individually and in a team, without bothering each other too much and with the possibility to have some private space to relax and find their own balance.

Living the *confinement* in small environments in which all the activities are concentrated in little space and in which one must live for a very long time (without the possibility of going out to take a breath of air, as it happens on board the ISS)

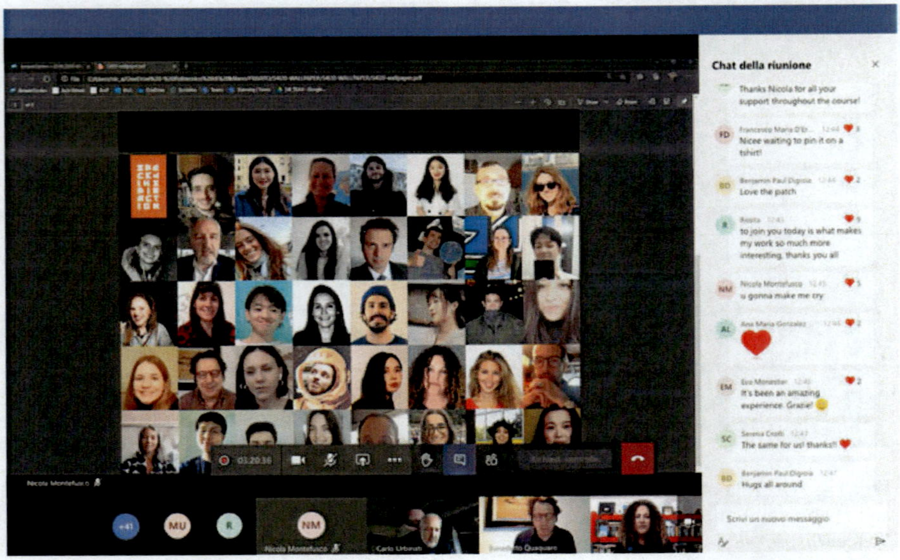

Fig. 1.5 Living in forced isolation during the Covid-19 pandemic and teaching in distance helped us to be more close to the astronauts' feelings and understand better their needs. A screenshot taken during the final presentation of *Space4InspirAction (S4I)* Course, 4th edition 2020, Politecnico di Milano. Credits by the author

causes an increase in the need for privacy, and cascading, emotional stress, aggression, nostalgia and depression. There are numerous tensions experimented on board between the crew, and between astronauts and ground controls, such as the mutiny of Carr, Gibson and Pogue against the mission leaders in Huston, ended with radio silence from the three men of the Skylab (Di Bernardo 1997).

Perhaps the closest experience to us that can be comparable to the life on board the ISS is the forced isolation, living in our homes, imposed by the Covid-19 pandemic emergency (Fig. 1.5). I had the occasion to live this experience with the students starting in the second semester this year, at the end of March 2020, the Space Design Course *Space4InspirAction (S4I)*[1] at the School of Design, Politecnico di Milano, that it is presented as case study in this book in chapter 4 *Space Design Between Research, Project and Education*. There are some specificities of Space Design that makes this discipline an interesting testing ground for

[1]*The Space4InspirAction (S4I)* Course, this year at its 4th edition—created and led by Annalisa Dominoni and Benedetto Quaquaro from 2017 inside the Master of Science in Integrated Product Design, School of Design, Politecnico di Milano—is the first and unique academic Course of *Space Design* in the world recognized and supported by the European Space Agency (ESA) through experts and scientists who suggest and then deepen the project themes, in line with the objectives of the all space agencies' strategic programs, with the aim to generate disruptive ideas and visions to increase the well-being of living in Space through Design approaches, methodologies and thinking.

Fig. 1.6 Desert RATS are led in the high-desert terrain and isolation of the Northern Arizona and are designed to exercise prototype planetary surface hardware and representative mission scenario operations focussed on equipment and spacesuit tests, vehicular excursions and exploration. Credits by NASA

the development of *distance learning* methods during a pandemic: astronauts live in a confined environment that brings lack of privacy, deprivation of natural stimuli and sometimes depression or nervousness for the impossibility of getting out that were close to the situation of us and our students *living in isolation*, maybe without a personal space in which to work or relax, and with the desire to move and practice sport in the open air. In this situation, students could be closer to their *users* and their *needs* to design new products which could better satisfy the functional, physiological and psychological requirements when living on board the ISS.

The designer can act to reduce the negative effects of confinement, designing an artificial environment rich of sensorial stimuli that are able to reproduce in the best possible way those already present in a natural environment.

The Design skills are also required to design the layout of the *human space settlements*, the architecture of the space stations, such as Mir and ISS, and the lunar and martian bases, the last two for now only projects on paper, and sometimes some habitat demonstrator units in 1:1 scale to test dimensions, habitation requirements and activities inside the base and outside the territory. These activities, as Desert Research and Technology Studies (Desert RATS or D-RATS) led by NASA, are designed to exercise prototype planetary surface hardware and representative mission scenario operations in relatively harsh climatic conditions where long-distance, multi-day traversing activities are achievable (Fig. 1.6). At this scale, from the lens of Design, the project develops focusing on the objectives of logistics, the functional distribution of indoor and outdoor spaces, as well as the *transition* of the crew passing from inside the habitat to outside the territory; on the applications of innovative technologies and on the most appropriate structural configurations to integrate the living areas to the surrounding territory, maintaining a systemic and organic vision of the project. The logistical nature prevails and requires of considering different alternatives to transport to Mars, for example, all the material needed to build the various modules of the settlements. The aspects linked to the dynamics of living, in which Design plays a fundamental role, are still far away. Today, the task of architects and designers is aimed at the development of structural concepts and feasibility studies of housing scenarios, in which the morphological and distributive aspects of the spaces are assumed. The most popular idea is to use the Moon as a test bed to gain experience in anticipation of the next missions to Mars, but also for other destinations, thanks to temporary habitats that integrate autochthonous materials and allow astronauts to live and work at the lunar base sheltered from the hostile environment.

Another case study described in this book in chapter 4 *Space Design Between Research, Project and Education* concerns the concept of a Moon base[2] called *Moony* built inside *lava tubes* using inflatable technologies for the habitat modular system, and 3D printing technology to create a safe rigid structure with *regolith*. Creating a permanent habitat on the Moon, precisely under its surface, in the *lava tubes*, is a challenging project at this moment. In our project, it is supposed the habitats will be pressurized and heated, with a useful area of about 125 m^2 and a height of about 3 m. They will require only three/four astronauts for assembly operations, which are not expected to last more than a few hours and can be disassembled and repositioned many times. It is important to keep in mind that the crew will work outside wearing the bulky suits for Extra-Vehicular Activities (EVA), which make everything slower and tiring, so the simplicity and speed of the habitation modules installation of modules are among the priority requirements. During the simulation period, a system of sensors will allow to monitor structural and housing performance to study how to optimize energy consumption, how to reduce packaging during transport, how to facilitate assembly and disassembly operations, and how to increase comfort for the life and work of the crew. The projects of interplanetary colonization, in which we work on the architecture, in the broadest sense of systemic organization, the human performances in relation to the environment and the tools that will be used on site, at the moment, increasing value. Even if at present, greater importance is focused on robotic assistance, especially in the early exploratory phases of the territory, human space missions are recognized essential to test the habitability conditions looking at the great opportunity to expand life on other planets.

---

[2]Moony—a whole modular base located under the lunar surface, in the natural tunnels of volcanic origin called *lava tubes*—was designed by Irene Zaccara and Emilia Rosselli Del Turco under the supervision of the professors Annalisa Dominoni and Benedetto Quaquaro, experts in Architecture and Design for Space, at the Design School of the Politecnico di Milano. The project is part of *Igluna Esa_Lab@* program supported by the European Space Agency (ESA) and coordinated by the Swiss Space Center.

Another Design field of research and project concerns the *clothing system* specific for extreme environment, with micro and reduced gravity, that consider the transition between inside the pressurized modules and outside the Space environment. The results of the project of the clothing system for Intra-Vehicular Activities (IVA) that I developed in two phases—*VEST, Development of an Integrated Clothing System*, and *GOAL, Garments of Orbital Activities in weightlessness*—and that I was able to test with astronauts through two experiments on board the International Space Station (ISS) during the missions *Marco Polo*, in 2002, and *Eneide*, in 2005, confirmed to me, but especially to the scientific community, that the designer really has the opportunity to increase comfort, psycho-physiological conditions, and the quality of life and work in Space: it is a pioneering field still little explored and, for this reason, with great potential for the contributions that the Design can give. The projects and the experiments *VEST* and *GOAL* are treated as case study in chapter 4 of this book *Space Design Between Research, Project and Education*. The implementation of these projects of garments designed specifically for Space—in which comfort, smart textile to increase body *hygiene and thermoregulation*, and *wearability* were for me the priorities to achieve—also provides researches and experimentations for new performance of fibres and fabrics that should have responded in an increasingly precise way to the environmental requirements in terms of *safety* and *toxicity*: encouraging, for example, the possibility to collect and contain, through the tissues in contact with the skin, the skin peeling, an important and amplified phenomenon in confined environments where hygiene and air quality are inevitably compromised; providing for the insertion of sensors in garments to monitor the biomedical parameters of the crew, without interfering with the subject or with his activity and without requiring direct and conscious participation; integrating to the clothes systems prosthetic wearable to support the movement and the stability of the astronauts; and considering different degrees of reduced gravity, so as to facilitate the maintenance of the posture during the various activities (Dominoni 2005).

I am leading in this moment researches on this topic, considering that the activity in Space is directing its attention towards planetary colonization programmes that include different environments and levels of gravity, in which astronauts will carry out activities in orbit and on lunar and martian territories—passing with a certain regularity from places internal and external—and must therefore have a clothing system adapted to the different environmental conditions. One of the most important aspects to ensure the protection and safety of human beings concerns clothing, a system of multi-layer garments able to ensure adaptability to various environments with reduced gravity—International Space Station (ISS), lunar orbital platform Gateway, Moon and Mars bases and external surfaces—and flexibility in movements for carrying out various Intra-Vehicular Activities (IVA) as work, fitness, rest, self-care, food preparation and consumption, etc., and Extra-Vehicular Activities (EVA) as rovers guide and robotic instrumentation, walks, surface missions for excavation work, extraction, construction, component assembly, etc. These researches are part of a wider, medium-term project to develop a new clothing and accessories system for both IVA and EVA consisting of multi-layer garments made of intelligent materials that integrate innovative technologies focused on health care and wellness, protection, and support for low-gravity human movement.

The Design domain fields for Space are very huge and we can say that embrace the entire artificial world created by the human beings and can therefore be of great help to suggest a more adherent approach to the needs of the human beings also in Space. Through the application of methodologies and techniques aimed at identifying factors as security, functionality, usability requirements and pleasantness, to the advantage of an increase in quality of life, not only psycho-physiological, but above all emotional, condition that will become more and more also important for space travellers. This is the reason why the designer has such a function important in Space Design, as extension of our world.

1.2.2 The Role of Space Design

Before going in depth into Space Design discipline, it is indispensable for me to focus on the meaning of Design and its role, which is something more and beyond the only shapes and surfaces of the objects. You know that today in Italy to define a beautiful object and that it has a style we use to say: it is "di design".[3] I do not mean it is little, since the *made in Italy* is recognized as excellence worldwide.

Actually, Design can play a much wider and more strategic role. I like to define Design approach as a *problem-creating* discipline, which generates innovation and increases people's well-being, and this is true both in Space and on Earth. The Design places the human beings at the centre, with a *User-Centred Design (UCD)* methodology, trying to interpret their needs in the most sustainable way: if an astronaut lives and works in a comfortable environment, with equipment and facilities designed properly to be used in Space, its performances increase as well as the quality of the activities, individual and collective, and can be determinant for the success of a mission. So, the most significant characteristic that distinguishes the designer's approach is the pre-eminent attention to the users of the products.

When I started designing and researching for Space more than 20 years ago, I wanted to go further to a *UCD* approach that belongs to Design discipline and find new tools specific for Space. In my first book *Industrial Design for Space*, as I already mentioned at the beginning of this chapter, my aims were to demonstrate through the results of the projects I developed during the first years the fundamental and strategic role of Space Design to increase the comfort and the well-being of the crew and, as a consequence, the good results of the whole human space mission.

[3]Even if it is inappropriate, a literal translation of "di design" in English should be "of design" that means an object made in Italy, which belongs to the *Italian design style*, an object beautiful and well done.

The structure of the book is composed of three principal parts: the first it is an overview of *Living in Space*, focussing on the principal characteristics that make Space an extreme environment, like *microgravity*, the biggest difference with Earth which causes strong alteration in the human being both physiological and postural (Fig. 1.7), the *confinement*, in which people have to live in small rooms in isolation for long time as already experimented in other extreme environments like Antarctic bases, and *cosmic radiations*, the most dangerous for the human being and whose effects are still little known; the second part, the core of the book, presents the potentialities of Design to have a central role in Space and describes a new approach *Designing for Space* bringing as case studies my first projects that shown the development of my researches and the results achieved; in the last part, I focus on the strategic objectives and on the new methodologies, as the *Use and Gesture Design (UGD)* that I found to increase the design process, according to the Space requirements and methods of working in Space, and the *vision* I wanted to exploit, in which the human being is at the centre of the human space flights, considering physiologic and psychological factors, together with functional and emotional needs.

I aimed to underline the peculiar condition in Space, among which microgravity and confinement are the major challenges to face with, and in which Design competences could offer valuable solutions to reduce difficulties of living and working in Space: for example, the attitude to consider the effects of microgravity not only on the physiology of the human body, but on objects and postures assumed by the human body, looking at the alteration of objects and movements with the environment, as well as the attention for the isolation issues related to spend long time in a confined environment, without privacy and the possibility to go out to take a breath (Fig. 1.8). My objectives were to report the theoretical reflections and the practical experiences acquired during 3 years of research activities, led for my Ph. D. in Industrial Design at Politecnico di Milano, focussing on the topics which arise from dealing with, and designing for an environment so unfamiliar as Space is.

I wanted to create a new discipline, nobody before has written a book on Space Design, but I did not want to limit myself to the description of what I might have done in this field, I wanted to bring real examples, case studies able to support my thoughts and demonstrate that Design could lead to disruptive ideas for Space.

I chose Space because it is the most complex and innovative field on the international scene, but overall, the most challenging, full of potential and still unexplored project opportunities, making it extremely interesting to the scientific community involved in industrial design. Therefore, I started my Ph.D. in 1998, the same year in which the construction of the International Space Station (ISS) began, and I considered this coincidence a very exciting and promising point of start. The conviction that animated me put Design as a bridge between Space and human being needs, able to dialogue with different communities of engineering, science and technology and give shape to new products and environments for the next generation of space travellers. Space today is an increasingly important and strategic field: the human beings are pioneers and are moved to the primal essential need to understand where we come from, and what surrounds us. It means

Fig. 1.7 In Space astronauts experiment, the *microgravity* which changes the physiology, the body posture and the relations with objects and environment. Credits by NASA

exploring new worlds, investigating the presence of life on other planets, and at the same time, caring for the Earth, protecting it from external, but also internal threats. For the same reason, we went beyond the oceans and the poles, towards unknown horizons, we are attracted by the Moon, to know it and discover, for example, how one can survive in its orbit for days. Only in this way we will be able to move towards Mars, overcoming all the critical issues of such a journey: the impossibility of being supported by Earth, the exploitation of in situ resources. None of this would be possible without accumulating experience and professionalism in the industry. We are called to understand what problems could be faced in the future as we are invited to anticipate them to solve them before they present themselves, and Design can play an important role in the whole process, thanks to its methodological approaches and strategic tools. The designer must interpret, by means of the formal and symbolic qualities of its projects, people's needs, expectations and desires, and at the same time elevating the degree of their general well-being. One of the visible signs that testify to the evolution of civilization is precisely the level of *comfort*—from the satisfaction of essential needs to their progressive sophistication—which is reflected directly by an improvement in the quality of life (Maldonado 1987). This explains why until now Space field has not included Design expertise in the strategic programmes of the human space missions: Space is an environment that has been subjected to very limited exploration, and in many ways, it is still an environment of pioneering, although great progress has been made since the 1950s. The boundaries of intervention areas for Design seem to

Fig. 1.8 The forced isolation in *confined environments* without privacy and natural stimuli alter physio-psychological factors and emotional needs. Credits by NASA

converge into a new figure specialized in the *new human factors* halfway between Design and human sciences (Jordan 2000) characterized by increasing attention towards the user and the value relative to the use and the emotions aroused by the product, on one hand, and by strong involvement in the processes of technical–scientific innovation on the other hand.

From the Design point of view, a good knowledge of the Space environment and, above all, of its effects on the human being is fundamental to identify the needs and the requirements, developing a path for the projects to follow, to offer the crew beneficial effects in terms of comfort and efficiency during their on board activities. But to gather information regarding life in Space, it is not an easy task: Space is an environment of pioneering with very limited access, manned missions are still few, and the experience collected by the Russian and American missions over the 60 years of activity has produced insufficient data to be transferred and translated into scientific knowledge that can now be shared. Also, being the de-briefing of each cosmonaut or astronaut subjective, the information about the experience of a human being in Space is sometimes vague and at times contradictory. To this lack of reliable information, I add the fact, proven by my direct experience while de-briefing with astronauts, that it is very difficult to obtain indications by them that could improve the efficiency or the quality of the existing products they use on the International Space Station (ISS).

Astronauts are trained to follow commands and procedures and maybe, for this reason, it is not easy for them to ask: *what if?* What if this tool had a different grip?

Could I improve the performance of a restraint? In which way? What would I anchor myself better? What would I need to move better?

Design is also asking the right questions and making the right choices according to the feedbacks collected in order to increase the well-being of the crew.

During this time, the expansion of manned missions in Space exploration and, above all, the gradual increase in the length of the uninterrupted periods spent by cosmonauts and astronauts, before on board the Mir and then the International Space Station (ISS), have encouraged researches and studies carried out on the habitability and focussed to confined environments in microgravity conditions.

The main objective is to guarantee now and in the near future—looking towards Moon and Mars as next interplanetary human settlements—a way of living and working which may be more comfortable, more functional and consequently more efficient: an expected first acknowledgment of the Design discipline for Space.

The growing attention given to human needs is also evident from the evolution of the strategic programmes for Space exploration led in the recent years by the National Aeronautics and Space Administration (NASA) in which the development and the importance of the *human factors* are strictly connected with the *security* of the human space flights. The consequently increased awareness of the crew's physiological and psychological needs that have always been neglected and considered secondary compared to other priorities, as typical of pioneering fields, has allowed that engineers, physicists, chemists, biologists and doctors to have with increasing frequently the collaboration of psychologists, sociologists, architects and designers, very useful for the definition of the context and the first requirements of a human space mission. It appears evident, therefore, how the importance of the *quality of life* is gradually spreading throughout most of the activities that involve the presence of humans in orbit, imposing itself strategically through projects aimed specifically at providing efficient *countermeasures*[4] to the disadvantages induced by the extraordinary situation concerning the life in Space. To this purpose, the study of the physiological behavioural and interpersonal dynamics, together with the interactions between the human being and the environment, can furnish instruments for the kind of investigation that is of primary importance during the design process and can lead to solutions really capable of improving the crew's level of well-being. Despite the obvious importance of Design to facilitate activities and life in Space, the scientific community is still dominated by an engineering matrix that fatigues, for the cultural education and background, to understand the contribution of Design for the success of a human space mission: I will never tire of repeating that if an astronaut feels better, thanks to a Design-oriented approach aimed at meeting the needs of the crew, also its performance and the quality of the results increase. But for now, as will be seen in chapter 3 of this book *Research and Design for Space Life*—in which I describe the main involvements of designers in projects and

[4]The Space scientific community uses the term *countermeasures* to describe all the devices, facilities, physical exercises, medications, drugs, procedures and other strategies that support and help astronauts healthy and productive during human space flights, with the aim to counteract the negative effects due principally to the microgravity.

researches for Space from the beginning of the *Space Race* until now—the complete acknowledgment of Space Design is still far away. But something fortunately is changing. The SpaceX's Crew Dragon Spaceship (Fig. 1.9) in partnership with NASA's Commercial Crew Program allowed through Design to add an emotional value to the human space flights, *making the experience of the crew more comfortable and beautiful as well as efficiency and safety* (Fig. 1.10).

One of the objectives that I would like to achieve with this my book is to give *visibility* to Space Design. In the scenario which I would describe, the Space field appears ready to accept and include, in the planning of the projects and definition of requirements, those factors that specifically pertain to Space Design, because of its highly multidisciplinary nature, can coordinate all the anthropologic, social, sensorial and perceptive aspects that regard the relations of the interface between operators, their second skin, the garments and tools with the surrounding environment.

The contribution of Design in Space, in this particular environment characterized by extreme conditions—three of which, microgravity, close confinement and cosmic radiations, have a determining effect on the human being and, as a consequence, on decisions regarding habitation projects—is to increase the comfort and the well-being of the crew designing instruments that can facilitate human movements and improve the execution of the various activities, to the advantage of greater productivity and a higher level of the quality of life in on board. I have to make it clear—after my more than 20-year experience in researching and designing habitation modules and tools that have been experimented and used by astronauts on board the International Space Station (ISS) and looking at the international experiences collected in few tens of years by space agencies and Industries—that increasing the well-being of the crew, the functionality of the activities and the usability of interface's instruments and tools for astronauts, we can achieve better goals that directly affect the success of a human space mission.

In this third millennium, the human beings are definitely the protagonists of a *New Space Era* and have at last conquered a primary role within the scale of values that characterizes a human space flight: they are no longer only professional and super-skilled astronauts, with years of demanding training in preparation for the mission, but can be also scientists, researchers, and not least, tourists ready to experiment space travels, but not trained therefore to live in an extreme environment, in conditions of permanent emergency, with limited resources and services, to the detriment of their physical and mental well-being.

The *New Human Being* is a person who wants to live in a comfortable way and, why not, has fun. This means that the whole environments must be totally re-thought and re-designed according to the physiological and psychological needs of the space travellers, as well as objects, facilities and tools for the daily life activities (living, working, sleeping, eating, taking care of ourselves, resting, etc.) which should make easier the life in Space trying to reduce the difference between living in Space and living on Earth. For these reasons, the role of Space Design is fundamental.

Fig. 1.9 SpaceX Crew Dragon Spaceship made history by docking to the ISS on 31 May 2020 carrying two NASA astronauts after 10 years from the last mission of the Space Shuttle. Credits by SpaceX

1.3 Use and Gesture Design

The lack of possibility to reference to a consolidated experience built over a very limited number of human space missions and the impossibility of assessing the true validity of the designer's hypotheses until it is already in orbit jeopardize the success of the project development in the Space field. For this reason, the visualization and analysis of all the gestures and movements possible in relationship to extra-terrestrial environments should be introduced as a determining factor during the design of the objects to deepen the dynamics of human body when in microgravity. A designer must have the capacity of imagine how could be the life in Space, also if it is not part of the normal experience of a human being, how his body could move in microgravity, how their postures and gestures would change in relation to objects and which could be the new objects to design properly for Space. *Designing for Space* means starting anew, applying a different logic for a different environment, conceiving new instruments for uses and activities difficulty in envisaging for us who are living on Earth, but which presuppose a different relationship between our bodies, the objects and the surrounding environment.

I found that the most important and most interesting difference between Design for Earth and Design for Space can be identified in the design process, which regards the capacity to *forecast* the design of use and gesture visualizing possible human gestures and movements in Space, in confined environments and in microgravity conditions, which cannot be compared to the human experience on Earth. This means imagine, for instance, which could be the alteration sustained by

Fig. 1.10 SpaceX Crew Dragon Spaceship interiors are designed to make the *experience more comfortable and beautiful*, maintaining as well as efficiency and security requirements. Credits by SpaceX

the astronauts assuming the Neutral Body Posture (NBP) and predict what the needs might be supported according to the ergonomic, physiological, perceptive, psychological, emotional, motor factors facing conditions completely new and unknown to the human being (Fig. 1.11). In order to manage better the transition between the two environment and practices, Design for Earth and Design for Space, I created a specific methodology which I called *Use and Gesture Design (UGD)* that means projecting the features of the future object on the scene of its possible uses, and visualize it in action: in your hands, put on you, in the environment to zero gravity where the object will be used, and foresee how it will be used, in which ways, in which environments and with which interfaces.

The design activity to create a new object, or an environment, for Space, should be contemporary to the forecast of the use of the new object, or environment. In this design process, the designer becomes a filmmaker, able to create a potential script of a new tool, which implies the relation with the environment, the movements and the gestures of the astronaut. It is as if the formal idea and performance of a project developed contemporary to a potential screenplay of movements and gestures of the operator (Rabardel 1997). In the design process, the artifact is of equal importance compared to the usage's scheme and creation of an idea, a concept design for the space, derives from the simultaneous design of action, movements and gestures that are simulated in function of how and could be used the designed object, how and with which procedures, how long and by how many operators, etc.

If *Designing for Space* means projecting the whole features of the new object on the scene of its possible uses and visualize it in action—in the hands of the user, or in the case of garments and accessories worn, and in the environment of

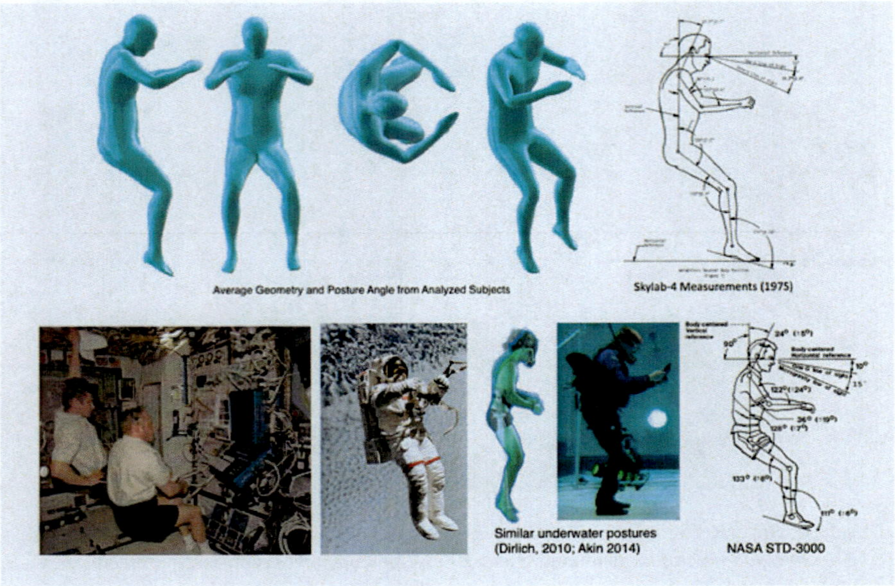

Fig. 1.11 Examples of Neutral Body Posture (NBP) in Space and similar underwater postures with average geometry and measurements. Credits by NASA-STD-3000

microgravity where it will be used—it follows that also the user is the artificer of the object, because its performance makes it complete and fully completed the project. Each use and gesture action has in fact to be understood as an *interpretative action*.

The transversal reference to the background of the choreographic ideology—which is to say the structuralist intention on the basis of which every action can be separated into a series of single movements and reconstructed as a sequence of movements—may be useful when illustrating the methodology that the designer should apply when designing for use and gesture. Like a director who must conjugate the engineering of screenplay narrative and the figurative nature of the storyboard—tracing the thread through the identification of the nodal points essential to film editing, and include and determine the geometry of actor and extra movements, as well as camera movements, connecting all with soundtrack rhythms—so the designer of structures and equipment for living and working in Space expresses, through a programme of use or a programme of gestures, a predefined sequence of actions that enable operators, in this case astronauts, to reconstruct designed movements for use of new products and therefore to reproduce them.

Going back to the choreographic ideology, the designer must not only foresee the series of actions involved in the use of the object, but must enhance, at the planning stage, both the gestures and the movements that accompany the object, and the object itself, like in a dance, must take on the role of the artifice of a series of gestures, a programme of ritualistic movements.

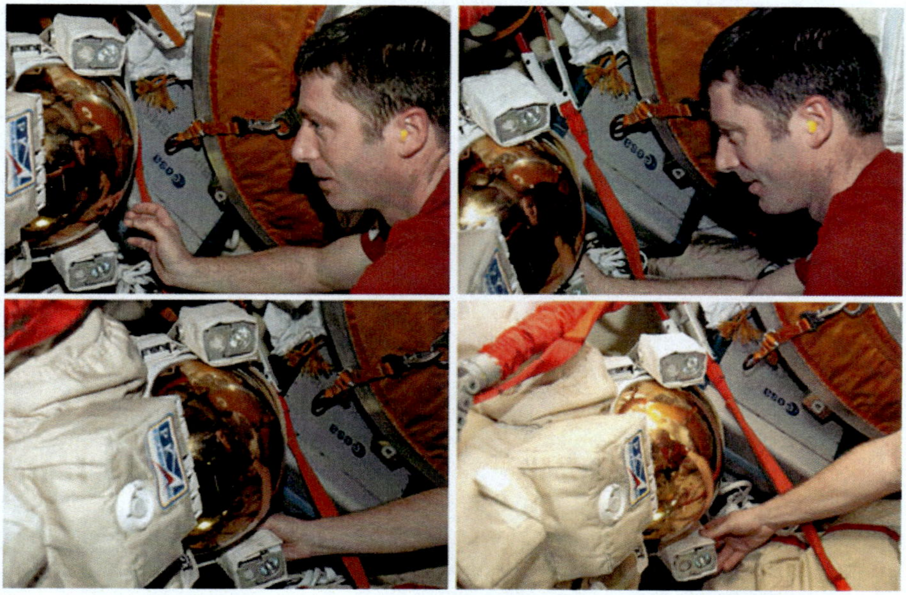

Fig. 1.12 A sequence of gestures and movements performed by the astronaut Roberto Vittori during *GOAL* experiment led by the author on board the ISS. Credits by the author

The first project on which I defined and applied the *Use and Gesture Design (UGD)* methodology in reality, realizing prototypes that have been tested underwater to check comfort and usability, was during the development of a system of portable tool holders that I designed for Thales Alenia Space (TAS) called *Portable Caddy System* in 1998. This device system was be carried or worn by the crew members for use as an aid in the ordinary maintenance on board the Columbus European Laboratory of the International Space Station (ISS).

Compared to the existing containers at that time on the ISS that are rigid toolboxes sized made to be integrated into the racks room dedicated to storage, I studied a system of soft containers, fabric, that was portable, and also wearable easily by astronauts. The project follows the principle of considering *microgravity as an advantage*, not a limit, trying to exploit the new potential that can have weightless objects floating in space. The idea was to imagine the containers close to the astronauts, as useful assistants, but independent, of the kind of *console* that could become rigid—if you had to lean and fix the tools that were not needed at that time, avoiding that they begin to float in the air, or that they were attached in a scattered way to the body of the astronauts with the Velcro®—and return soft, to be rolled, once used the tools and stored in the *Portable Caddy System*, to occupy the minimum of space possible. This project is described in detail in chapter 3 *Research and Design for Space Life* at the paragraph *The Birth of a New Discipline*. The Design is based on the possibility of simulating on Earth the operator's movements by means of advance verification of the body's possible uses and gestures in relation to the objects and the Space environment. Design in compliance with use and gestures—that is, the analysis of the movements most suitable for using the containers and the tools and recomposed them into precise sequences of gestures—was carried out inter-linking it with the parallel development of the morphologic and functional characteristics of the tool holder system. To plan the gesture of *how* to use the containers and the tools—that is, how to recognize the tool required for a particular operation, how to take it out of the container, how to anchor it to the body and to the structure of the Space habitation module in a situation of microgravity, how to put it back into the container, etc.—and to design the shape of the objects themselves—relative to the physiological, perceptive, ergonomic, psychological and motor aspects of the operator, but also to the structural, material, mechanical and functional aspects of the instruments on board—are actions that interpret a unique design process where theory and practice overlap, and in which the value relative to the use of an object is strictly connected to the intrinsic value of the object itself. To have recourse to consolidate practice is desirable both for the designer, who can operate in a familiar field, according to predetermined and consequently easier to manage conventions, and for users, who have the advantage of greater confidence in and control of actions, which are the same as, or similar to, familiar gestures.

Forecasting capacity is one of the requirements most important that a designer must possess for design gestures for objects and propose innovative solutions that are not polluted by experience acquired of daily life, but which can instead respond to latent and not yet expressed needs in a conscious way by the people: functional needs and emotional in which the poetics of the object has a fundamental role.

This skill of having a strong capacity to foresee the new designed objects and environments together with the use and gestures that will be performed by the astronauts (Fig. 1.12) is strongly connected to another one that embraces *cognitive psychology* and the mental processes behaviour to design products and create environments well organized (Norman 1988) and adapted to the needs of people, in order to find solutions really capable of improving the crew's level of comfort and efficiency and the whole well-being on board.

Recently, I developed a design research exploring architecture of motion as scenarios of movement and stasis, very perceptible in Space, to evolve the *Use and Gesture Design (UGD)* methodology integrating experiences of *kinaesthetic awareness* and generate new ideas and scenarios to facilitate creativity and innovation (Dominoni and Fairburn 2015). We believe that the level of consciousness coming from the awareness of body, object and environment, of each gesture and movement necessary to perform a task, can help the designer and the *performer* to analyse in depth the need and the most hidden behaviours of people. Looking at Space as inspiration we wanted to offer scenarios that may be of use to enrich design processes for non-extreme environments. Working and observing astronauts who live in microgravity conditions we have been *forced* to slow down the rhythms to which we are accustomed on Earth during the design process. Could slowness become a new and important approach in the design process to innovate? In this context, the Design becomes a *slow mediator* able to read usages, gestures and behaviours of the people and translate them into new useful products to increase the well-being and answer to the need of the humans by observing the experience of living and working in Space or on Earth, and also, by transferring the results from one field to another and vice versa. Could a constructed design scenario, for example, an immersive experience in an unfamiliar environment, induce a state that shifts our perspectives and in turn generates new conditions for Design approach?

1.4 Designing the Space Experience

These themes refer directly to the *User Experience Design (UXD)*, the process aimed at increasing users' satisfaction by improving usability, ease of use and overall the pleasure provided in the interaction between the user and the products. The design of the user experience includes the traditional human–machine interaction design and extends it to all aspects of the product or service as perceived by users. Pushing us further, we take into account that today not only trained astronauts, but also researchers and tourists look to Space as an opportunity ever closer and possible. And so, why not also think about *designing the experience* of those who want to make a *spacewalk* in all safety and freedom (Fig. 1.13), with a drone following them, which is always close to them, and then "embraces" them to bring them back into the International Space Station (ISS)? Of course, the same drone, in addition to serve and entertain space travellers, can serve astronauts during Extra-Vehicular Activity (EVA) and, therefore, could assist them during difficult and dangerous operations outside the ISS, where time spent in Space is inversely proportional to the security of the mission. This project, an example of *User Experience Design (UXD)* applied to Space, was developed inside the Course of Space Design named *Space4InspirAction (S4I)* that I created and direct, together with Benedetto Quaquaro, at the School of Design, Politecnico di Milano, inside the Master of Science in Integrated Product Design.[5] *S4I*, treated in this book as a case study in chapter 4 *Space Design Between Research, Project and Education*, is the first and unique Space Design Course in the world recognized and supported by the European Space Agency (ESA) with experts and scientists who suggest and then deepen the project themes, chosen every year together with us, in line with the main objectives of the space agencies and industries' strategic interplanetary programmes. The collaboration between our *S4I* Course and ESA is not a spot project, but the beginning of an ongoing agreement about *teaching—research—design* to ensure the development of innovative projects in which the contribution of Design discipline is fundamental and strategic to generate disruptive ideas and visions for habitability in Space, with particular attention to human factors and interactions between operator, tools and environment.

[5]While I'm writing this book, we're leading the 4th edition 2020 of our Course of Space Design *Space4InspirAction* in distance learning for the Covid-19.

Fig. 1.13 A *spacewalk* in totally freedom and security, thanks to a drone-mate designed to increase the value of the experience for astronauts and space travellers. Credits by *Space4InspirAction (S4I)* Course, 1st edition 2017, Politecnico di Milano

Aim of our Course S4I is to create new professional figures able to connect both technology and beauty languages, increase creativity and visioning, find design solutions crossing know-how and research, imagine new cultural and business models to increase well-being according to sustainability.

The Course *Space4InspirAction* is for the students an opportunity to develop a strong capacity of visioning looking at Space as an inspiration, to innovate and design new projects of space habitats, objects and tools, but also new scenarios that could help people to live better in a sustainable way (Dominoni et al. 2018). The Space field allows them to develop creativity with intense visioning activities by confronting with confined and microgravity environment that is not part of our common experience. Thanks to ESA support, we are able to compare our design vision with real conditions and requirements and find innovative solutions designed for Space, but that can be also transferred to daily life on Earth becoming spin-offs, and vice versa, becoming spin-ins.

In the case study of the *spacewalk* (Fig. 1.13), I showed you an example of how Space Design can design new exciting experiences for both astronauts and tourists, creating equipment and objects the support their exploration of Space, but we could go further. The project of the drone is useful for me to introduce another potential strategic role of Space Design which concerns the ability to find new applications of Space technologies and behaviours to generate spin-offs and spin-ins, from Space to Earth and vice versa. The drone for the *spacewalk* may be a suggestion for another

Fig. 1.14 The drone for the *spacewalk* may be a spin-off for another extreme environment, but on Earth, which could perform diving safer. Credits by *Space4InspirAction (S4I)* Course, 1st edition 2017, Politecnico di Milano

extreme environment, but on Earth, which could perform diving safer (Fig. 1.14). This project can be considered a spin-off from Space to Earth, and an example of our methodology of designing through the crossover of Space and Earth, along with different disciplines and areas of applications, to bring two seemingly incompatible environments closer because they are both inhabited by human beings. And the role of Design is to look at the human being increasing the comfort and the well-being of the people for a more sustainable society. I believe Space innovations will have a strong influence on how people behave and perform in the near future, as well as Design that gives shape to our world and tries to humanize technology to meet the needs of a sustainable society. I also believe Space Design discipline can really inspire students—it is enough to think the strong attraction that Space has always generated on society, thanks to its narrative power—and help them to understand better how they can be more conscious of the transformation technology brings.

1.5 Design for the New Space Economy

The Space designer could be seen as a contemporary *influencer* of our society and the Design discipline a booster for the new Space Economy.

Designing for Space, and specifically with microgravity conditions, extends the area of design research outside of the common rules and consolidate methodologies that we use to consider and apply in a design process. Looking through the transformative lens of Space, we are freed from the conventional references and potentially advantaged by different points of view and new scenarios (Dominoni and Fairburn 2015) and at the same time, we have the advantage to apply the lessons learned from one extreme scenario to another one. Using the experience of Design for Space to inform methodologies for Design on Earth may generate a paradox, as clearly microgravity isn't a condition on Earth, but therein lies the opportunity to find analogous ways to create this state, and other states, to generate new conditions, in turn leading to design innovation.

The fascination for Space started in the 1960s during the *Space Age* while there was an optimistic mood for the conquest of Space, brought with it a great expectation of growth and trust in the future. Design looked at Space as inspiration to create many products became *icons of style* that deeply influenced the aesthetics of our world. Today, in the era of the Space Economy, in which finance and private companies have become the boosters of the new Space Race, Design become a *producer* of ideas and suggestions, a *connector* between futuristic visions and technological development. So nowadays also the role of Design has changed and no longer deals only with giving a *space mood* to the objects: Design is a method of actions, a *problem-creating* approach that *drives innovation through research* and develops business success exploring and understanding the users and their needs in a sustainable way.

1.5.1 Design Research for Innovation

Research for innovation represents today the industry's indispensable tool to hinder the international concurrence throughout new Design solutions aimed to optimize the know-how and the skills of each single firm. The enormous change and the exponential speed of transformation driving the evolution of contemporary society aims the designer to understand how to manage complexity and find solutions to the problems arising from it.

Likewise, industry is forced to undertake design research for the only fundamental resource that allows it to create a competitive advantage to weather market turbulence: knowledge. If we think about the most significant innovation factors of company systems which are spreading in the market, and the strategies suggested by the management, it seems evident that many methodologies applied inside the companies belong to the theory and practice of the Design discipline.

A design research, to be considered such, must produce some form of knowledge that should be useful, but also lend itself to being shared with the scientific community and the industry within which it is generated and to being extended to the outside world as well. That is to say, we could sum up the notion of design research with the formula: *Design + Knowledge*. The design research is a programme of research and action characterized by quality which defines itself gradually studying and negotiating constantly oneself reasons with the aim to focus the attention on the context and the situation without punctual analysis or précised collected data (Laurel 2003). This means to measure oneself throughout our own personal wisdom in front of a reality unforeseeable and in continuous movement interpreting the results of data in a subjective way. Let us deepen into Design and innovation relationships.

Design is to formulate hypotheses, identify ideal solutions, outline future scenarios and choose. It also means receiving stimuli and processing them, confronting oneself with models of thought and value systems. Which happens, above all, when referring to a line of antecedents, that is turning to the past logical context of the problem one wishes to solve. This leads us to the conceptual relationship of planning with creating, inventing and discovering and, last but not least, deciding. According to Tomás Maldonado, the creative moment is to a large extent identified with the all but irrational one of problem-solving (Maldonado 1999). And it is here that one discovers what inventing and discovering have in common with Design, because *Design always means confronting and creating problems*. And nowadays, the size of the problems to be confronted and created is such that it is bringing about a change of project activity and, indirectly, of the role and tasks of the designer. The approach of professional people who work in the Design field has more and more in common with that of researchers who, when confronted with complex problems, are willing to collect information, organize it so it is ready for use, formulate hypotheses and come up with innovative solutions. Observing reality and people behaviour, generating ideas through brainstorming process, mocking up working models in order to visualize possible solutions, refining concepts and creating product or service with technological and scientific implementation are the techniques more utilized by the best design bureaux, as IDEO, to make *good design* (Nussbaum 2004). The interest of the whole system of the business companies for the design research brings to the absorption of the Design strategies in the managerial sphere producing a hybridization between design and management culture and the emerging of the design thinking as core of the new decisional process. And the design thinking is based on some specific capabilities of the design strategies which are connoted by visual and perceptive aptitudes. The design research is an articulated whole consisting of the expertise, capacity, sensitivity and tools used by the *designer-researcher* to deal with complexity and the pace at which said problems change so as to then translate such design experience into new knowledge that can be transferred and acknowledged by the scientific community in which the research itself was generated. This sort of figure, somehow embodying the relationship between research and practice in industrial design, can operate in universities as well as with firms having state-of-the-art research and development centres. To work on design research, one needs a real principal submitting a problem and the conditions characterizing it, in short providing a context: the principal would thus come to engage in constructive dialogue with the designer to generate innovation.

Innovation today is no longer only a single enterprise initiative, but an innovative result born from a network of relationships and cooperation that comes from the outside and even through the organization of *multidisciplinary working groups*, capable of contributing with other points of view, to the growth of a new manufacturing reality (Dominoni 2009). Innovation is a system of activities and behaviours aimed at an objective to produce growth: human resources, know-how, financial skills, technological tools, but overall, research and development thought connections and networks with different subjects and realities.

Design and innovation tend to overlap when we find that Design is synonymous of innovation centred on the human needs. And the collaboration among Design, industry and science is the base for successful projects, products and services. Nearby the expressive, formal and functional qualities of industrial products, the Design has been enriched of knowledge which belong to engineering, technology and innovation economy. Also, the research activities are articulated and extended beyond the boundaries traced out at the beginning from the discipline, through unexplored territories of investigation and experimentation, increasing new Design questions and new tools to answer them. *Design is distinguished from the other disciplines for a strong attitude in regarding beyond the functional aspects of the products with the aim to generate new visions, concentrating on "how" innovation can find "potential applications" in different fields, allowing spin-offs of technologies, materials, processes and working methods which can evolve in new products and services systems*. To innovate is necessary to consider some actors that emerge strongly from the world global change which impose companies' new *Design tools* to confront the market dynamics and their effects on human beings, society and the environment.

The base of this approach suggests to give a great attention to *balance*, during the design development, four important actors, with the aim to consider and frame innovation opportunities in a way that leads to success in the market: the *companies' needs*, to explore new productive and business potentialities in relation to the market transition maintaining the pre-existing know-how, with the aim to increase the brand value; the *users*, to understand which products could reflect human lifestyles, values and aspirations in order to absorb people outside our existing market; the *technology*, to find the best technologic tools and processes compared with the other three actors; the *environment*, to preserve our natural resources for a more sustainable future (Brown 2009). These actors have to be considered inside a strategic macro-area in which the design capabilities are integrated with the management structure of the industrial system to produce innovation in terms of new processes, products and materials crossing different expertise and areas of application to generate spin-offs and spin-ins: towards a new design culture that goes beyond Earth into Space.

1.5.2 Space Inspires Innovation Through Design

A new*Space Era* is slowly coming into focus. It is a future where space agencies won't be the only actors and will increasingly expand their partnerships with the private sector (Hufenbach 2017). The International Space Station (ISS) is the fastest moving incubation centre, and it has proven itself as a business accelerator: scientists and private companies are invited to take advantage of this opportunity. I was invited by the European Space Agency (ESA)—which organized a new kind of event in London during the autumn 2016—together with Benedetto Quaquaro to discuss cultural aspects of Space introducing some of our works, as *Fashion in Orbit* that used Space as inspiration.[6] The *Space for Inspiration—ISS and Beyond* event (Fig. 1.15) was an open invitation to a broader community to join the Space club, and it was also a unique opportunity to build cross-sector relationships and learn about promising research taking place in orbit. ESA invited Space and non-Space industry representatives to sit together for the first time and exchange ideas on what the future of Space exploration holds for them both on Earth and in Space.

The newly emerging discipline of *Space Design* is in between Space and Earth that combines technological research, space inspiration and aesthetics to enhance the research of behaviour for life in Outer Space and on Earth. The role of Space Design is strategic because it is able to speak both the different languages of science and beauty putting at the same time the human being in the centre. If astronauts live better in a comfortable environment (considering functional, physiological, psychological and emotional factors), their performances and the success of the whole mission increase. This is the reason why Space Design can increase the comfort of the astronauts and find new applications for space technology in daily life on Earth. In this context, the Design approach can be a powerful engine for the Space Economy able to translate Space inspiration into new products involving both Space and non-Space industries.

Space exploration offers fantastic scenarios of experimentation that can be transformed and reinterpreted by the language of Design, an area that is far removed from the scientific world and yet has an important role to play by finding new products or applications for Space technologies that can enhance the performance of people and the quality of life (Dominoni and Quaquaro 2017a; b).

Furthermore, the new Space Economy is looking for attractive ways of communicating Space. In an age crowded with digital innovation mainly involving the movements of our fingers, the Space community must find a new tool to entice the public and help explain why Space research and its impact are so important: for this

---

[6]*Fashion in Orbit* is the first Higher Education Course in Space Fashion and Technology created and directed by Annalisa Dominoni and Benedetto Quaquaro at POLI.design (the Consortium of Politecnico di Milano) as consequence of the success of *Couture in Orbit* project and supported by the European Space Agency (ESA). *Couture in Orbit* and *Fashion in Orbit* are presented as case studies in chapter 4 of this book *Space Design Between Research, Project and Education*.

Fig. 1.15 Benedetto Quaquaro and I (second and third from right to left) at *Space for Inspiration—ISS and Beyond* event organized by the European Space Agency (ESA). Credits by ESA

reason, the Design approach can really be the new *space tool*, and I explain here why through two projects I developed recently.

The first project, *Couture in Orbit*, is an example of spin-off: we have been inspired by space technology to find new application in terrestrial products to be used on Earth increasing the garments' performances.

The second project, a *Space Acoustic Object*, is an example of spin-in: we transferred innovation technologies used in new products on Earth to Space environment with the aim to reduce noise on board the ISS.

1.5.3 Space Technology Spin-Offs Through Fashion

The project *Couture in Orbit* was launched by the European Space Agency (ESA) in 2016 to communicate to a huge public the role of ESA in developing innovation and *inspiring people through design and fashion language* that celebrates space exploration and research exploitation, and their impact in many fields of our life. I was invited by ESA to represent the Italian side of the project to create a *capsule collection* of garments inspired by space scenarios, using ESA technologies, and designed to be worn every day on Earth. The results were presented with a catwalk[7] of the whole collection at the London Science Museum. This project is treated as a case study in this book in chapter 4 *Space Design Between*

[7]The video of the final catwalk of *Couture in Orbit: from Spacewalk to Catwalk* at the London Science Museum, European Space Agency (ESA), 28 September 2016 https://www.youtube.com/watch?v=B99ncZ5eSM0&list=ULtD4c8evE4YA&index=42&app=desktop.

Fig. 1.16 The Higher Education Course *Fashion in Orbit* can be considered a spin-off of *Couture in Orbit* project, supported by ESA and POLI.design. Credits by POLI.design

Research, Project and Education. For the moment, it is useful to me to introduce *Couture in Orbit* here, and the following spin-off *Fashion in Orbit* (Fig. 1.16).

My principal aim was to find relations between life in Space and life on Earth connecting the two environments and considering not only technological spin-offs or spin-ins but also emotional language and usability.

Together with a group of students—that have been chosen specifically for this project from the Master of Science in Design for the Fashion System which I led at the School of Design, Politecnico di Milano—we transformed inspirations into shapes and textures, by looking at astronaut postures and movements in micro-gravity and considering their activities in Space, in relation to the objects they use, to generate new gestures and ideas for our projects. I wanted to imagine new uses of space technologies that they do not yet have applications other than the original, but they could in a short time become very interesting.

Translating high fashion concepts into the language of technology was an essential skill for an endeavour like the project *Couture in Orbit*. I have been the go-between, with the students on one side and the industries involved by ESA on another, among which Rina Consulting, the Italian partner of ESA for *Technology Transfer Solutions*, and Extreme Materials, a small company specialized in textile-related technologies, which has in the past included such innovations smart materials embedded in textiles and composites. The students have sought a clear explanation of how these technologies are used in practice, and in other cases, consulted with the companies involved in the project, to find original solutions and developments to the problems they have posed.

Fig. 1.17 The *rose* on the knee recalls a real configuration of an ESA patent for a 3D antenna which we imagined printed on textile with conductive ink. This project, a garment of our *Couture in Orbit* collection, is an example of how Design can add an *aesthetic value* to technology. Credits by Politecnico di Milano, ph. M. Bergamini

1.5.4 Adding an Aesthetic Value to Technology

We pushed creativity and encouraged the students to come up with apparently crazy applications—to stay as crazy as possible—while we were the ones with our feet on the ground, to take into consideration the feasibility of implementation of ideas in concrete, workable way.[8] For instance, could a wearable cooling system be modified to gradually deliver fragrances instead? And could an innovative ESA patent for a 3D antenna can be printed directly on textiles with conductive ink? The drawing similar to a rose made of concentric little circles that you can see at the knee height (Fig. 1.17) is the configuration of a real antenna created by ESA which currently consists of three-dimensional cones, and that we imagined printed on fabric with conductive ink.

You can foresee the potential applications of this idea on clothing and accessories to amplify the signal of the mobile phone in reception and transmission. But you can also imagine how design can add an *aesthetic value* to technology. Paradoxically, the idea of the dress is disruptive even without the technology working. In addition to the configuration of the ESA antenna, which is objectively a beautiful pattern, we were inspired by *Space Age* of 1960s choosing colours like silver and white in all the lunar nuances, looking at materials such as aluminium and reflective surfaces, selecting curved shapes and imagining textiles printed with wonderful pictures taken by astronauts with their camera, looking out of the

[8]https://www.esa.int/Science_Exploration/Human_and_Robotic_Exploration/Couture_in_orbit/The_go-between.

Fig. 1.18 This *Space Acoustic Object* realized by Annalisa Dominoni and Benedetto Quaquaro for Caimi Brevetti combines technological research, Space inspiration and aesthetics showing that Design can be a *bridge* between human sciences, technology and beauty. Credits by Caimi Brevetti

International Space Station (ISS) towards Earth, as well as by satellites, with marvellous colours, patterns and phenomena of light.

In the era of the Space Economy, the designer becomes a *producer* of ideas and suggestions, a *connector* between futuristic scenarios and technological development. Though design is possible to create a dialogue between distant productive knowledge to generate innovation, find new applications for space technologies which can be integrated into new *terrestrial* products to improve the quality of our everyday life and vice versa, to involve companies of the private sector which do not belong to the Space field, but that for their know-how could make a great contribution to the development of human space flights.

1.5.5 Space Technology Spin-Ins Through Design

The second project, a *Space Acoustic Object* (Fig. 1.18), can be considered an *icon* of our way of designing between Space and Earth, a project that combines technological research, Space inspiration and aesthetics crossing different fields of applications: this product increases astronauts' well-being, because it is made of specific material able to absorb the noise when living on board the ISS and creates a dialogue between distant productive knowledge, because of the contamination through distant knowledge, to generate disruptive innovation.

This *Space Acoustic Object*, which I designed together with Benedetto Quaquaro[9] for Caimi Brevetti, was born by combining the sound-absorbing technology of *Snowsound Fibre*[10] fabrics with the textile processing of luxury fashion, giving rise to a rather surprising shape, suggested by the structure of the pleating, which, in addition to be beautiful, increases the acoustic performances due to the compression of the fibres. It consists of a frame in painted tubular steel, a core in fibrous sound-absorbing polyester material and a cover in plissé *Snowsound Fibre* sound-absorbing fabric, as well as in polyester. Physically large, it can be fixed to a wall or ceiling where it creates a dialogue between different production techniques: the sound-absorbing technology of *Snowsound Fibre* fabrics blends with the textile processing of the fashion industry.

Meanwhile, with Thales Alenia Space, we are testing new acoustic products to use them on the International Space Station (ISS). *Pinna*[11] is a concrete example of how Design can speak different languages and act as bridge between human sciences, technology and beauty.

1.5.6 A Bridge Between Human Sciences, Technology and Beauty

In the first chapter, I presented the strategic role of Design for Space underlining its capacity to create disruptive ideas and to be a bridge between human sciences, technology and beauty to generate innovation and increase the people's wellness, in Space and on Earth, for a more sustainable society.

Design can speak different languages, combines science and beauty, becomes a mediator between space technology and needs through projects that generate new scenarios and new paradigms with which to confront. We have seen in the previous paragraph some examples of what Design can do for Space and that the role of Design for the space industry derives from the need to carry out research in confined spaces and microgravity conditions, performed by human beings, in order to test out new scientific equipment and investigate environmental and behavioural

[9]Co-owner of the design studio (a + b) Dominoni, Quaquaro, founded in 2011 in Milan by Annalisa Dominoni and Benedetto Quaquaro with the aim of contaminating research and technological innovation with poetry. Among their latest projects, in addition to a feasibility study of *Preliminary Design Solutions for the Support to New Orbital Infrastructure Recreational/Habitable Configuration* of a new space station for Thales Alenia Space, there are the *Leonardo* trains of the Milan Metro for ATM, produced by Hitachi Rail, and the *Space Acoustic Object* for Caimi Brevetti, both selected by ADI Design Index.

[10]The patented Snowsound Fibre technology by the design company Caimi Brevetti is based on soft interwoven polyester acoustic fibres that are inherently fire-resistant. The interaction between the special acoustic fibres and the design of the object and the support system helps reduce annoying acoustic reverberation in rooms, while improving quality of both life and work.

[11]We named *Pinna* this *Space Acoustic Object* because of its shape: a large three-dimensional fan that resembles the fin of a fish.

dynamics in relation to the crews. We know that we need to increase know-how and experience to design inhabitable systems for long interplanetary voyages and missions, to colonize other planets, starting from Moon and Mars. Design can play a crucial role in making everyday activities much easier in any setting or context, helping us perform more efficiently and also enhancing our mental and physical well-being. The Design discipline studies surrounding reality and interprets all those signals we receive from the environment by focussing on how people live, behave, move, show their emotions and interact with each other. A designer is capable of leaping *out-of-the-box* and developing a new and different line of thinking translating latent demands, not yet expressed by the people, into projects designed to help astronauts, serving emotional as well as practical purposes.

The task of Design in Space is to be a bridge between science and culture of beauty, finding a common language that can help the hi-tech research and industrial worlds communicate and collaborate. When I tried to define the boundaries of the Design discipline in this first chapter, I affirmed that designers are no longer problem-solvers, asked to come up with solutions, but they are rather *problem creators* capable of calling into question the world in which we live, exciting and amazing the general public just like artists can do, thanks to their ability to explore the complex and delicate process of interaction between people, objects and environments (Dominoni 2015). If on one side Space Design can increase the comfort and the efficiency when living and working in reduced gravity, exploring and understanding the astronauts and their needs to find innovative solutions to make easier human performances on board, on the other side, it finds new applications for space technology, adding also an aesthetic value to the final project, and drives the development of business models. I would like to conclude this first chapter with a comparison between a picture of the interior of the International Space Station (ISS) as it is today and an image of how it could be the next interior of a new spaceship, looking at example to possible solution for the cis-lunar orbital station Gateway that we have developed inside the 2nd edition 2018 of the *Space4InspirAction (S4I)* Course of Space Design at the School of Design, Politecnico di Milano, supported by ESA and Thales Alenia Space.

On board the ISS space is reduced to a minimum, each task provides procedures to conduct experiments, but not to adapt optimally to the confined environment, astronauts must adapt along with direct experience in the field. The visual perception of the whole space is quite chaotic, with cables and devices put everywhere, that makes tasks of the crew arduous and more difficult to achieve (Fig. 1.19).

The interfaces of the equipment are for the greater part obsolete if we consider the advances of the technology and above all the development of solutions intuitive and more *friendly* in order to facilitate the use of many today household appliances of wide consumption on Earth. The experiments, prepared by scientists and engineers who do not bother to make the interface human-friendly, are quite rudimentary "boxes" that must work. Since it is no longer possible to do without machines, we should learn to live with them and communicate with them, *imagine a technology with a soul* (Norman 2007). Space is a pioneering sector and should be at the forefront in investigating and experiencing the most innovative interaction

Fig. 1.19 The astronaut Nicole Scott installs hardware in the Fluid Integrated Rack in the Destiny Lab in which the chaotic environment makes her task more arduous and difficult to decodify. Credits by NASA

dynamics. Instead, the crew lives and works in an environment that is the opposite of what it should be: comfortable, orderly and efficient way. In my opinion, the Space Design's contribute on the environment would be able to improve the living and working conditions of the astronauts, focusing on the well-being and on the psychological and emotional aspects that, together with biomedical ones, are essential for the success of a mission. The internal volumes of the Space habitation modules should be completely re-designed, maybe alternating large areas with foldable structures that could be transformed according to the various needs of space; mobile and light tools could be fixed at any point of the Space Station to work or rest; interactive screens could cover the instrumentation, while allowing astronauts to interact with them, to achieve an overall visual order and a more clear perception of the whole environment.

What about a new *sensitive environment* for Space (Fig. 1.20) in which interactions between the astronauts, equipment, machines and spaces are intuitive and requires simple tasks? This idea, described in chapter 4 of this book *Space Design Between Research, Project and Education,* inside the case study of *Space4InspirAction* Course, is about a specific design concept for new astronauts' private spaces. By using *soft robots* and flexible OLED technology, we could be able to create a comfortable personal space where astronauts can relax, enjoy their free time and sleep. The *soft robots* are now able to expand themselves six times the starting length giving the possibility to assure astronauts a reactive product that is also able to provide pressure on their body and making him feel more comfortable with microgravity consequences. This is just an example that looks forward the future technology, then slowly but inevitably *biological technology* will arrive,

Fig. 1.20 What about a new *sensitive environment* for Space in which interactions between the astronauts, equipment, machines and spaces are intuitive and requires simple tasks? Credits by *Space4InspirAction (S4I)* Course, 2nd edition 2018, Politecnico di Milano

perhaps accompanied by devices for perception, memory and even for the increase of muscle strength. We are living now the expansion of an augmented and immersive VR to allow a wider and varied perception of space. Designers are the first to raise the problem because they are the ones to translate ideas into reality and to produce impacts on society.

References

Brown T (2009) Change by design. How design thinking transforms organizations and inspires innovation. Harper Collins Publishers, New York

Clark P (1988) The Soviet manned space program. An illustrated history of the men, the missions, and the spacecraft. Orion Books, London

Di Bernardo G (1997) Vedo la terra azzurra. Editalia, Rome

Dominoni A (2002) Industrial design for space. Silvana Editoriale, Milan

Dominoni A (2005) VEST. Clothing support system on-orbit validation. In: SAE technical paper series, ISSN 0184-7191, Proceedings of ICES 35th international conference on environmental systems, Rome, 11–14 July 2005. SAE Society of Automotive Engineers, Warrendale, Pennsylvania, USA

Dominoni A (2009) Creative network system for innovation. In: Proceedings of INTED 2009. International technology, education and development conference, Valencia, 09–11 Mar 2009. Association of Technology, Education and Development (IATED), pp 002572–002580

Dominoni A (2015) For designers with their head beyond the clouds. Politecnica. Architettura, Ingegneria, Scienze, Maggioli Editore, Milan

Dominoni A, Fairburn S (2015) Designing from the unfamiliar: how designing for space and extreme environments can generate spin-offs and innovative product. In: The value of design research. Proceedings of EAD 11th European academy of design conference, Paris Descartes University, 22–24 Apr 2015

Dominoni A, Quaquaro B (2017a) Dressing up for space. ROOM. Space J 13, Fall 2017:106–109. The Aerospace International Research Center (AIRC). Publishers Press, Vienna

Dominoni A, Quaquaro B (2017b) Future fashion. Space research inspires innovation in fashion. ROOM. Space J 10, Winter 2017:102–107. The Aerospace International Research Center (AIRC). Publishers Press, Vienna

Dominoni A, Quaquaro B, Pappalardo R (2018) Space design learning. An innovative approach of space education through design. In: Proceedings of IAC 69th international astronautical congress, Bremen, 01–05 Oct 2018

Hufenbach B (2017) Space is open for business. Room. Space J 10, Winter 2017:102–107. The Aerospace International Research Center (AIRC). Publishers Press, Vienna

Jordan P (2000) Designing pleasurable products. An introduction to the new human factors. Taylor & Francis, London

Kline P (1995) Models and personality traits in occupational psychological testing. Int J Sel Assess 3:186–190. https://doi.org/10.1111/j.1468-2389.1995.tb00026.x

Laurel B (2003) Design research: methods and perspectives. MIT Press

Maldonado T (1999) Tecnica e società. I nuovi scenari. In: Riccini R (ed) Il Contributo del disegno industriale, All'insegna del Pesce d'Oro, Milan

Maldonado T (1987) Il futuro della modernità. Feltrinelli, Milan

Norman D (1988) The psychology of everyday things. Basic Books Publisher, New York

Norman D (2007) The design of future things. Basic Book Publisher, New York

Nussbaum B (2004) The power of design. Business Week, May

Rabardel P (1997) Gli strumenti dell'uomo. Dal progetto all'uso. In: Ergonomia, n. 7

Souza K, Etheridge G, Callahan PX (1991–1998) Life into space. Space life sciences experiments. Ames Research Center, Kennedy Space Center 1991–1998

Stuster J (1986) Space station habitability. Recommendations based on a systematic comparative analysis of analogous conditions. NASA Contractor Report 3943. Scientific and Technical Information Branch

UNESCO (2015) UNESCO science report: Towards 2030. UNESCO Publishing, Paris

Chapter 2
Living in Space by the Lens of Design

2.1 The Transition to Microgravity

On board the International Space Station (ISS) the interior of the habitation modules, in origin a cylindrical housing, is *covered* by racks—the modular storage system which contains all the experiments, facilities and equipment useful to the maintenance of the spaceship—and there is never enough room in a confined environment: for these reasons, astronauts live in an extremely chaotic space (Fig. 2.1) very difficult to encode and where it is a *challenge* to maintain the sense of orientation, because on ISS we do not have up and down, unless by convention (Taylor et al. 1987). They take 15 days to orient themselves in labyrinth paths that connect the various modules *working with terrestrial instruments and objects which are adapted for use in Space* such as machines and tools for physical exercises, or utensils for food preparation, consumption and conservation. They have *facilities* for personal hygiene with an "interface" for their needs not properly comfortable, not to mention all the physiological, psychological and perceptive alterations that the human being suffers in a confined environment with reduced gravity (Souza et al.1991–1998).

In human space flights, there is a substantial change in posture and in the dimensions of the human body due to the effect of microgravity. The new posture is similar to the one assumed by the body underwater, called by NASA Neutral Body Posture (NBP), and seems to result principally from a new equilibrium of the muscular forces in relation to the tension of the tissues acting on the various joints. NBP has been adopted for newer system designs, including the computer consoles and space suits. NASA Human Factors and Health Technical Standards (NASA-STD-3000) specifies NBP as a reference posture of the human body in 0-g when relaxed and no external forces are applied. Maintaining a body posture other than NBP requires significant strength exertions (Stuster 1986).

A. Dominoni, *Design of Supporting Systems for Life in Outer Space*, Research for Development, https://doi.org/10.1007/978-3-030-60942-9_2

Fig. 2.1 On board the ISS it is a challenge to maintain the sense of orientation because there is not up and down, unless by convention, and the chaotic environment aggravated the perception of space. Credits by NASA

If relaxed, the body assumes a stooping position, in which the angle of the elbows and knees is about 130°, the pelvic angle changes, and the curve of the lumbar and thoracic sections of the spinal column straightens out, causing an elongation of the body of more than 10 cm.

The head and the spinal column bend forwards and the upper limbs float 450° upward towards the trunk (Wichman and Donaldson 1996). While the NBP in Space has no countereffects in itself, except for the discomfort that can be experienced when trying to reproduce terrestrial postures, like standing up or lying down, the implications of the new physical relationship between the user and tools must be considered carefully. A significant example is the design of the workstation: astronauts have difficulty in working at waist height, since they must continually exert an effort on the arms to bring them to this lower level.

The Design must therefore evaluate and verify the most suitable postures and movements of a particular activity and forecast the use and the gestures of the astronauts in relation to the interface of the instruments and the surrounding environment (Dominoni 2015). From another point of view, working in microgravity can allow to type contemporary on many laptops at the same time while they are floating in front of you (Fig. 2.2).

In addition to the new NBP, in Space the movements necessary to go from one point to another, as well as exert enough force to take an object in microgravity,

Fig. 2.2 From another point of view, working in microgravity can allow to type contemporary on many laptops at the same time while they are floating in front of you. Credits by NASA

require the adaptation to the proprioceptive[1] parameters, and it becomes very difficult to maintain one's balance and sense of direction. The proprioception is the sense of perception, usually at a subconscious level, of the movements and position of the body and especially its limbs, independent of vision; this sense is gained primarily from input from sensory nerve terminals in muscles and tendons (muscle spindles) and the fibrous capsule of joints combined with input from the vestibular apparatus. Besides the perception of spaces, volumes and colours, it is different and consequently, operating efficiency and performances during the various human activities decrease. Not only does microgravity have a determining influence on a human being, but also on the *force weight of objects*, increasing the difficulty in using and controlling them. The use of Velcro® is still currently the most efficient and the least invasive system useful to fix objects and tools to the interior surfaces of the spaceship, or the bodies of the astronauts, mainly thanks to its simplicity and its minimum volume and weight. "When an object is placed on a horizontal surface on Earth it stays there, held down by its own weight and by surface friction. In a spaceship, in the absence of weight, at first, an object left alone seems to remain

[1]The sense of *proprioception* is the ability to perceive and recognize the position of one's body in space and the state of contraction of one's muscles. It allows us to know by looking at a bottle, how much strength we will have to exert on it to lift it. Considering the weight in Space is near zero, all the forces exerted on objects and the environment must be recalibrated in order to move us, and our body can learn it only through direct experience.

innocently where it is—apparently giving you time to do something else and then find it again—waiting faithfully, without even having to lay it down somewhere. In fact, there is always a minimum residual speed, perhaps because of re-circulating air, which causes the object to move slowly and inexorably on its own account. If it is a small object, you end up losing track of it, sometimes finding it again by chance after looking for in vain. And sometimes it doesn't turn up again until after the return to Earth. This explains why almost all the objects on board, from pencils to manuals, spoons and cameras, have one or more Velcro patches. It has to be possible to attach every tool on board to Velcro® patches on the walls. Unfortunately, there is never enough Velcro on the walls, or at least not in the places preferred by the people on board. However, Velcro® is a plastic fibre that emits toxic fumes in the case of fire, and so the interior designers of the Shuttle tend to use only as much as is strictly necessary. There are spare strips of self-adhesive Velcro® on board and every crew ends up making abundant use of them. After its returns from the mission, every Shuttle is cleaned and invariably, the next crew will have to begin their Velcro wall papering from scratch" (Malerba 1993). Franco Malerba is not the only astronaut to be surprised by the behaviour of objects in orbit. Also, Paolo Nespoli told me that after returning from space missions on Earth it took a long time before he stopped breaking glasses that fell down while he expected that once let go after use, they would remain floating in the air. Once you change a habit, it's hard to let it go. For the same reason, astronauts must be helped and supported when they return to Earth because otherwise they would not stand, even after only 2 weeks in orbit. But not for muscle mass loss and bone decalcification that are important but not so devastating alterations in the short run. When the astronauts arrive on board the ISS *they must learn to recalibrate the weight force they are used to exert on objects to take them*, just looking at them, without making excessive movements that would be necessary to take the same objects on Earth, because in Space, objects maintain the same mass but are weightless. In Space, the lack of gravity produces jokes to our proprioceptive system and the strength that we usually use unconsciously to lift a bottle is much greater than necessary, and the result is that we risk launching it instead of lifting it. The astronauts will then have to gradually get used to recalibrating all the forces, this time consciously, to avoid inconvenience troubles. For the same reason, when they return to Earth after 6 months of life in Space they cannot stand up and walk because their body no longer remembers how much strength must exert on the muscles of the legs to counter gravity and stay standing (Dominoni 2002). Considering the weight in Space is near zero, all the forces exerted on objects and the environment must be recalibrated in order to move us, and our body can learn it only through direct experience. This observation is very important from a Design perceptive, because when we design an object to be used by an astronaut we have to *imagine uses and gestures of the body in relation to the object and the environment in microgravity*, in order to transform the microgravity, today considered one of the great limitations on the human activities, in an advantage, in an opportunity. Going further, as designers we should try to *enhance the sense of perception in Space* with artificial stimuli able to gradually accustom astronauts to these new parameters and, in the same way, reconnect them to the terrestrial proprioception parameters of their own, when they must return to the Earth. Even though with the lens of Design we can *transform microgravity in an advantage,* it is considered the greatest limitation and the most difficult problem to solve from the physiological point of view, at least for the time being, because of the harmful effects on the nervous system, the cardiovascular system, and the muscular and skeletal systems in particular.

Microgravity also causes variations in the dimensions and the morphology of the human body: the length of the spinal column increases, the circumference of the legs decreases and body fluids flow from the lower part of the body to the upper part, causing the typical *puffy face* phenomenon. The whole process determines a complete change in the morphology of the human body (Antonutto and Di Prampero 1992). The principal negative effects of microgravity on the human body that can be verified immediately are vertigo, a decrease in the quantity of blood, a horizontal shifting of the position of the heart and a decrease in its size, a loss of muscle tone and of muscle mass, bone decalcification, and decreased respiratory capacity, as the organs no longer being subjected to any gravitational pull drawn down, push towards the diaphragm and limit its movement. The reason is that the human cardiovascular system is structured for correct functioning in an environment of lG, contrasting the force of gravity by a force exerted in the opposite direction, which redistributes the body fluids in the upper part of the body. This system also works in the same way in Space in conditions of microgravity, causing a reduction in the volume of fluids in the legs and a consequential accumulation of these fluids in the head, neck and chest (Gazenko et al. 1982). The redistribution of fluids has important side effects. A system of interior sensors recognizes the excess of blood in the upper part of the body and it compensates by reducing the production of red blood cells. Consequently, after around 100 days spent on board, the astronaut begins to suffer from anemia. The same system of interior sensors also registers the excess of fluids in the upper part of the body and again attempts to rebalance this by reducing the desire to drink (Moore et al. 1996). The absence of the sensation of thirst in turn reduces the level of fluid in the body, causing bone demineralization and increasing sodium retention.

2.2 Countermeasures

The term *countermeasures* in Space includes all the aids—activities, equipment, instrumentations, outfitting, drugs therapies and physical exercises—useful to counteract the negative effects of the microgravity on the human body. Our body is not made for microgravity (almost 0 gravity)! We have seen in the previous paragraph how to do actions to contrast the bad effects: we have to adapt the body to the inhospitable environment; we cannot stand tall because of the Neutral Body Posture (NBP); we always must remember that it is important to stay fit in case of emergency; we do not know what happens to the body measures if we do not adopt countermeasures; more fluids towards the brain (*puffy face*); sweating? A bit

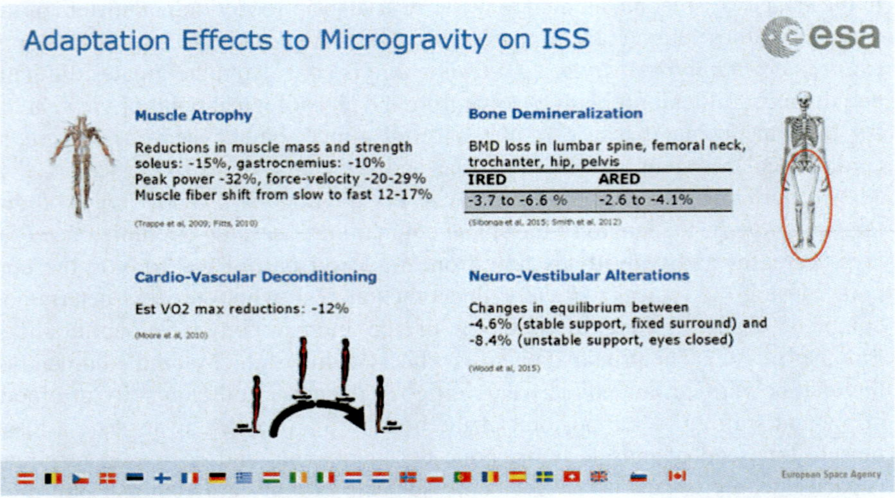

Fig. 2.3 The major effects of the microgravity on the human physiology during human space flights. Credits by ESA

different that on Earth with a bad feeling as "sticking bubbles on the skin" that they have to clean only with wet wipes.

If we go deeply into a scientific evaluation of the effects of microgravity on the human body we face a (Fig. 2.3) muscle atrophy, in which the more problems are related to legs (very critical) than on arms; cardiovascular deconditioning; bone demineralization, especially legs and pelvis, while arms are usually active; neuro-vestibular alterations connected with the sense of balance, humans are among mammals that can stay on two feet and this requires lots of coordination and balance. In microgravity, we cannot walk on two feet, so we lose the equilibrium.

The goals of the countermeasures are principal to make enable the astronauts to return safe on Earth in the same conditions as before the human space flight. The three major life-threatening hazards that can occur in a vehicle for Space are fire, depressurization and toxic atmosphere.

There are physical aspects of the human space flights that impact to the performances on board, and it is important that astronauts arrive prepared doing a lot of training activities before flight to face with: centrifuge training, survival training, system training, fit-check, frequent travelling, emergency drills, piloting and physical training. The countermeasure training is the hardest physical activity required to astronauts in Space, together with Extra-Vehicular Activity (EVA) preparation.

Physical exercises are considered the best countermeasure to counteract impact on physical performance capacity. In order to maintain muscle tone, at least two hours a day of intense physical activity are scheduled, to be carried out during the period dedicated to the daily on board working activities. Exercises must be

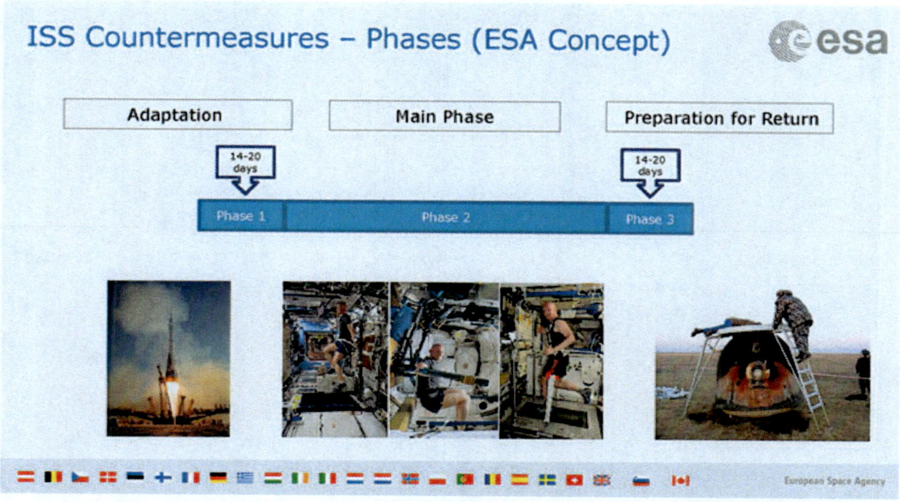

Fig. 2.4 There are three phases foreseen by ESA protocol to schedule physical exercises based on the approach "Pre-In-Post-Flight". Credits by ESA

selected in order to avoid body adaptation and its negative effects and an intense training is required to astronauts to generate multiple stimulations.

Fitness for Space is necessary for many reasons to maintain the workforce, enable a safe return, but also a potential emergency, enable human force on planetary explorations, enable quick recovery and adaptation to gravity environment, prevent injuries, as well as long-term health problems.

The physical training is foreseen with a step approach "Pre-In-Post-Flight":

- Pre Flight: Status quo, optimally prepare body.
- In-Flight: At the moment of the launch the body conditions of astronauts are at the top level, during the flight they have a remote support and a basis of papers to tell them which exercises they have to do. Two or three days after they arrive on board, they start with the exercise, even if at the beginning they have to adjust to the new environment, and it takes time.
- Post-Flight: They have to perform two hours of exercises on the ground for 3 weeks. Often astronauts are supported remotely, because they travel a lot before their missions so they cannot have personal trainers each day.

The strategies of exercises are aimed to produce a stimulation of muscles to avoid atrophy, provide axial loading to substitute gravity and training of prime movers. The state of the art in Space reports principally three phases (Fig. 2.4):

1. an "adaptation" to the extreme environment that takes more or less 14–20 days if considering the astronauts feel and not heavy loads;
2. a "main phase" in which the training intensity increase according to the body challenge to be sure it does not *relax* until their maximum possible extent;

Fig. 2.5 In the Russian habitation module of ISS cosmonauts perform training with BD-2 treadmill and VELO cycle ergometer. Credits by NASA

3. a "preparation for return" the same period of 14–20 days to intensify the training a bit more, a last little sprint to get ready to return to Earth gravity. Sometimes, the phases 2 and 3 are quite similar because phase 2 is already very intense.

Since strength and resistance are fundamental for carrying out the various activities in Space, the criterion used for selecting and training astronauts is an elevated aerobic capacity and muscular strength, especially regarding the upper part of the body. However, even those astronauts who have greater than average physical capacities, like marathon runners and weightlifters, inevitably lose a significant percentage of their strength.

The training devices currently used on the International Space Station (ISS) are Russian (Roscosmos), American (NASA) and European (ESA).

In the Russian habitation module of the ISS, we find the BD-2, a sort of treadmill that keeps the body in the right position, thanks to restraints to reproduce the artificial load that we have on Earth when running. BD-2 name comes from Russian *Begushaya Doroshka*, which it can be literally translated as the "Running Path", a Russian term for a treadmill; the VELO, a cycle ergometer (there are various types), has the principal purpose to keep muscle trophism as normal as possible (Fig. 2.5). During the space long-term missions on board the Mir, the Russian space station, the crew had to spend at least two hours a day doing physical exercises: an hour on the VELO ergometric cycling machine, with a variable resistance between 100 and

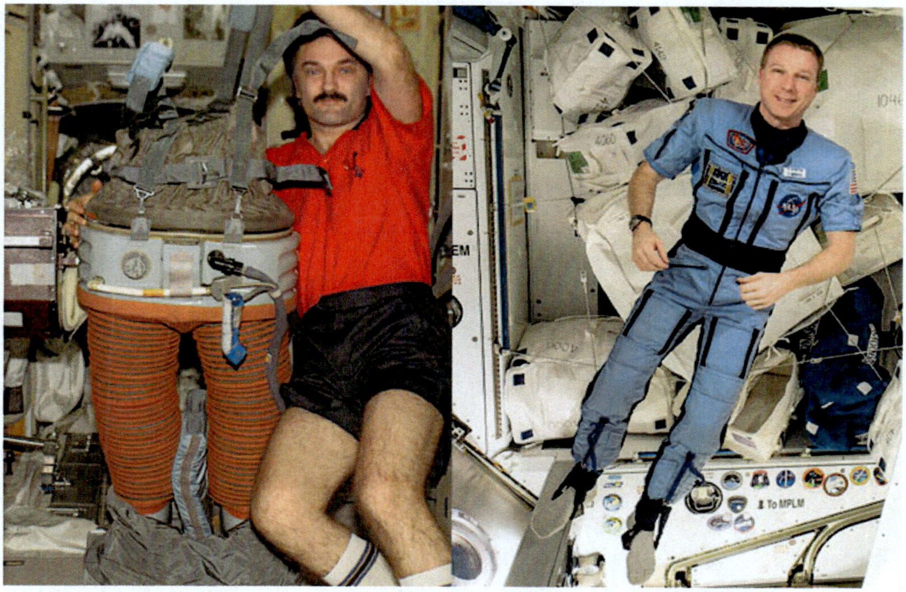

Fig. 2.6 The Low Body Negative Pressure (LBNP) balances the blood circulation system, and the *Penguin* simulates the force of gravity with elasticized bands. Credits by NASA

170 W that the cosmonaut himself decided, and an hour running on the BD-2 treadmill, to which the body was bound by means of elastic straps, exerting a force equal to around 50% of his body weight on Earth. Periodically, ergometers and force gauges were used which permitted physical exercise of the upper limbs.

The crew members also undergo the Low Body Negative Pressure (LBNP), used before going back to Earth, creating a negative pressure pulling blood towards the feet for the cardiovascular system balance by means of special rigid trousers with a negative pressure called "Chibis Suit", which cover the lower part of the body up to the waist, and a decompression that created a reduction in the pressure of 20–30 mm Hg below atmospheric pressure was created by means of a system of pumps (Fig. 2.6 left). This solution is to encourage the flow of fluids in the lower part of the body, partially reproducing the pressure conditions of the blood circulation system of a subject standing upright in the same force of gravity as on Earth. The aim of the LBNP is to reduce the deconditioning of the cardiovascular system that occurs during long human space flights.

The cosmonauts also use a suit called *Penguin*, which can simulate the gravity on Earth, weighing down on the spinal column and the muscles (Fig. 2.6). This suit is made of a very thick and elasticized fabric to generate muscle tension, while inside it has a system of adjustable rubber bands, from the shoulder to the hips, which exert pressure on the spinal column. It is a sort of scuba diving suit. The cosmonaut who wears it, for a minimum of 8 h to a maximum of 16 h a day, is obliged to make a continual effort to move, and since this demands extra effort from

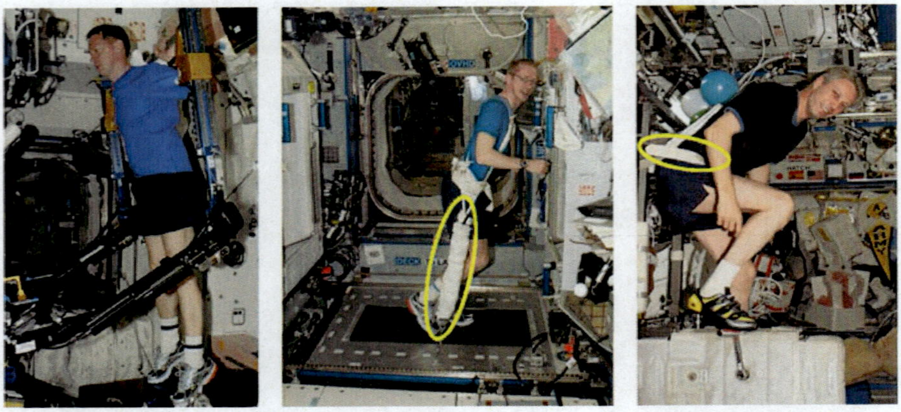

Fig. 2.7 ISS countermeasures training hardware adopted by NASA: ARED on the left, T2 in the centre, CEVIS on the right. Credits by NASA

the body in order to execute the tasks, he is in fact in a situation of constant physical exercise. The *Penguin* is a very good concept, but it is not very used because of the discomfort.

The American system foresees three devices as countermeasures on ISS: Advanced Resistance Exercise Device (ARED) was launched to ISS in 2008 and accommodated additional exercise protocols and had almost twice the loading capability of SpiraFlex Interim Resistive Exercise Device (iRED), launched in 2000 with the first ISS crew; T2 treadmill (like BD-2 for Russian); and Cycle Ergometer with Vibration Isolation (CEVIS), which is similar to a *cyclette* with click pedals and handle to stay in the right position (Fig. 2.7).

It is useful to see, *by the lens of Design* in order to improve the qualities and the performances of the equipment, how these tools have evolved throughout the years. iRED came before ARED and the maximum load switched from 136 to 272 kg. Why is there such a difference between the load of the two equipment pre- and post-ARED? Squat and Bench Press have been considered on both the equipment, but if with iRED the only way astronauts had to push the body to limits was making more repetition because the fitness machine was not able to provide more loading even if astronauts were able to reach higher loads, with ARED they can add a maximum load higher to make muscle force increase, otherwise they are not performing what they need to. ARED is used mainly for Squat. When we usually do exercises on Earth, the spinal column is fixed and related, while in Space it seems to have a curve in contrast with the movement of the machine. This happens because the muscles around spinal column decrease quickly for the lack of gravity, so the machine provides astronauts to move the column to avoid atrophy to those muscles which are more reduced. In fact, astronauts in the past had problems with their columns. Considering that the body in Space has to learn specific movements and there is no way to create them perfectly as on Earth with the support of devices, only Squat simulation seems to be close to the gym performed

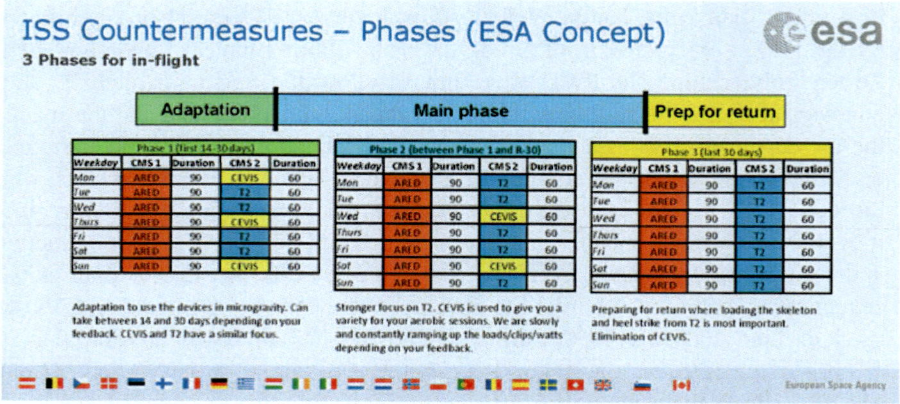

Fig. 2.8 The use of each device is scheduled to obtain the maximum results according to the three phases for in-flight: adaptation, main phase, preparation for return. Credits by ESA

on Earth. Astronauts move the machine with their movements, so coordination and rhythm are required.

Space constraints and *vibrations* are very important to consider. The cardiovascular training, which is performed with the T2 treadmill and the CEVIS cycle ergometer, produces a lot of vibration when in use. So, the power of the exercises must be reduced, for example, the maximum speed of the T2 treadmill cannot be so high in Space because the reaction vibration system must counter react to the running performance that is not like running on Earth with 100% of our body weight. We are never able to fully load the body like in gravity conditions. The base of ARED moves because of the microgravity: thanks to the isolation system, the forces are not transmitted to the Spacecraft and through the machine the data are given to Earth.

In this chart (Fig. 2.8), we can see when and how much each of these devices is used by the crew members on board the ISS: usually two hours a day, before it was one hour in the morning and one hour in the afternoon, but considering that the equipment and the area have to be cleaned each time after use, the crew prefer only one unique session longer. Also, the scheduled use of the exercises' machines must be organized since they are shared tools.

The European Space Agency (ESA) facility Muscle Atrophy Research and Exercise System (MARES) is integrated into the Columbus Laboratory of the ISS and is part of the NASA Human Research Facility. MARES was launched in 2010 as a result of researches into the effects caused by weightlessness on human muscle-skeletal system in the microgravity environment. The system enables the subject to carry out physical workouts with various groups of muscles of the upper limbs, lower limbs and of the trunk: bending/stretching of the ankles, elbows, knees, hips, trunk, abduction/adduction and rotation of the shoulder. Moreover, the apparatus allows so-called "polyarticular movements" like the bending/stretching of the lower limbs. The MARES is the natural evolution of a previously used instrument, called the Torque Velocity Dynamometer (TVD), also commissioned by the ESA, constructed in the mid-90s in the Biomechanical Laboratory of the Zurich Polytechnic. The TVD was built to allow the bending/stretching of the elbows and knees, keeping both the force opposing the movement and the speed of the movement itself constant. In addition, the TVD allowed both concentric exercises, in which the muscles contract against a force, and exocentric exercises, where the contracted muscle is stretched by an external force exerted against it. In all movements, special sensors monitored the force and the speed of the contractions, and by means of a special software programme it was possible to plan suitable sequences of exercises, according to the choice of the person using it. The TVD was used in June and July 1996 during the NASA human space mission *Life and Microgravity Science*, on board the Shuttle Columbia, which lasted 17 days. MARES allows the same type of uses as the TVD, obviously extended to a greater number of muscle groups.

In microgravity conditions, the human being is in a sort of *natural stretching permanent state*, sometimes they get muscles soreness, but not so often.

When astronauts use the cycle ergometer feet are fixed and they do not have the settle because of microgravity and must stay in position by attaching to a bar or a belt. It is the astronaut that fixes himself to the structure and this need to stabilize the body creates an extra consumption of energy, so we do not have to add loads because *the effort is in staying in position*.

Regarding testing in Space, it is very difficult and the only test we can assure is the one on the cycle ergometer that requires a lot of time and resources, considering that there are not fixed standard, yet the astronauts have to reach, developed monthly for 10–15 min by Private Exercise Conference (PEC). ARED foreseen 45 min of privatized audio session, 3 for mission and works in this way: the remote control centre see the machine and the astronaut on it, and people talk and help to him controlling the movements and support him in order to make good exercises. So, the feedbacks can be traced back to a coach who follows him while he is working. It is important to notice that data protection of the astronauts is directly connected with the privacy rights: they may desire to ask for medical personal information or discuss other private items that they do not want share with the other members of the crew on board. In the future, we expect to have new systems that provide feedbacks to the ground without real people connected in remote modality to increase the autonomy of the crew that could have instantaneous feedbacks. It is a great challenge, because we need big data, but it is crucial for future Mars expeditions.

Despite the big scientific research behind every project of a fitness machine for Space environment, together with the enormous number of people involved with different expertise and knowledge, it happens sometimes that the devices proposed to fail. It is the case of the YOYO™, Clinical Application of Iso-Inertial Eccentric Overload, that allows to perform exercise using spinning flywheels installed on the International Space Station in 2009. The method, applicable to any muscle group, provides accommodated resistance and optimal muscle loading through the full range of motion of concentric actions, and brief episodes of eccentric overload. So, apparently the concept was very good, but performing exercises it was found out that it could be too hard for wrists and in the final analysis very dangerous for the crew. The wrists worked instead of the legs, and this was an enormous mistake for the design of the machine. From this lesson, we can conclude that designing fitness machines for Space, but also for Earth, we always must think of which muscles are involved in the activity we imagine performing with the new tools.

To conclude this paragraph, it is very important to know for a designer that there are many operational constraints to consider to design new fitness equipment for Space: *Vehicle Constraints* that includes space availability, energy resources, air cooling, CO2 removing capacity, food and water supplies, thermal regulation, and body moisture; *Vibration Isolation Constraints (VIS)* discussed in the previous pages; *Schedule Constraints*; *Exercise Hardware Constraints*; *Individual Spread Between Astronauts* that concerns personal experiences and preferences; and *Multilateral Communication* that implies communicating with others improving the collectivity on board.

When we know the constraints, we can be more creative and at the same time propose ideas more feasible, surely disruptive projects could include lower mass and less space, micromovements performed with small tools, considering that balance and coordination must be improved and that some tools could be the same for Pre-In-Post-Flight.

I imagine also that as designers we should create new tools in the middle between *physical exercise and entertainment* to make more pleasant the fitness activity that today for astronauts is always very boring, as well as to allow performing exercises in couple, exploiting microgravity like an advantage, or in group, increasing the sense of belonging one community.

The new developments are concentrated to both the scenarios of Artemis Program—the US government-funded human space flight that has the goal of landing “the first woman and the next man on the Moon by 2024—and the Gateway —the new lunar orbital space station—that will be much smaller than the ISS and for that reason it will require smaller exercise tools than. A new multi-function fitness machine called the European Enhanced Exploration Exercise Device (E4D) is being developed by the danish aerospace company on behalf of the European Space Agency (ESA). A flight model is expected to be ready for use during human space flight missions in 2021. The E4D machine combines cycling, rowing, rope pulling and more than 29 other weight training exercises in a single machine. This provides more versatility and thus also enhances the effectiveness of the astronauts’ daily workout in Space. The goal that these fitness machines should share energy with the rest of the space station is not yet achieved, but it is definitely something useful to consider for Mars missions, and to include in the next Design.

2.3 Confined Environment

Living in *isolation* for long time produces a quantity of negative effects on the human physiology and psychology that impact on the quality of life and at the end on the general well-being of the crew. The well-being of the astronauts depends on a series of *environmental stimuli* such as light, colours, variations in the air, wind, heat, cold, smells, tastes, etc. which are normally present on Earth and which are able to activate vital body functions. The human being reacts to these stimulations by tuning his or her equilibrium to nature and to the surrounding environment, and the lack of even one of these stimuli can jeopardize the whole biological and mental health of the individual. Natural daylight, for example, has a direct influence on the *circadian rhythms*: in a closed environment, these rhythms undergo strong alterations and thus cause problems of insomnia and general intolerance, compromising the health and the efficiency of the astronauts in the performing their activities on board as well as the good results of the mission. Inside the ISS, for the moment, the artificial environment recreated inside the pressurized modules is unable to reproduce all the stimuli present in the terrestrial environment and, in the long term, this provokes a sense of monotony and a general lack of interest among the astronauts. The psychophysical state of the crew members, who become more and more unmotivated, tends to reduce vital functions to a minimum, even if sometimes they try to find personal solutions to counteract this problem, as the initiative of the cosmonauts Anatoly Berezovoj described in the first chapter (Clark 1988).

The confined environment usually implies the crew to live a small space for long time in which more people have to share everything, *no privacy at all*, without the possibility to go out for a walk. It is sufficient to consider that the standardized dimensions of the living quarters on the ISS have an internal diameter of 4.6 m which—considering the space taken up by the four *racks*, the modular replaceable "cupboards" which contain equipment, provisions, experiments and space for storage and which are arranged around the cylindrical wall surface of the module forming a square section—is reduced to only 2.2 m that can effectively be used by the astronauts (Peercy et al. 1985a, b).

The reaction to this extraordinary situation is the increasement of emotional stress, intensified by the fact of living in a *permanent state of emergency* that can give rise to feeling of anger and aggression. Astronauts are trained to manage that, but unpredictable situations can happen, and there is a *gap between to be prepared to a mission and living and working on ISS*. The crew must be able to check the aggressiveness that develops because of the lack of space, the forced isolation and the constant sensation of risk, a situation that is potentially aggravated by the heterosexual and multicultural characters of the subjects involved. Considerable tension has been recorded on board among crew members (Raasch et al. 1985).

A part the little space available, the isolation and the consequent *nostalgia* is the problem that generates the most stress and depression, even if the communications technologies and social media allow the astronauts to remain in contact with their families, and with the whole planet, in the most natural way possible.

2.4 On Board Astronauts' Activities

The necessity of a Space Design approach, in which the attention is oriented towards the user, is much more important and desirable in Space where, at present, inside the habitation modules of ISS equipped with technologically advanced apparatus, there still exist structures and equipment for astronauts that do not seem to take into due account the crew's physiological and psychological needs. Take, for example, the basic functions of eating, sleeping or taking care of personal hygiene which concern the human being, to name but a few, and verify how in the living quarters, the instruments and the objects necessary for living and working in Space have been designed until now (Raasch et al. 1985).

Eating is a primary necessity of the human being as well as sleeping or taking care of the personal hygiene. During the whole human space missions performed so far—including the actual ISS activities—astronauts could not cook: the complete absence of gravity and the impossibility to use stove or flames not allow them to prepare their own food. Despite this, experts of the space agencies are fully aware that "like on Earth, nutrition plays a key role in maintaining the health and optimal performance of astronauts before, during and after human space flight" (Douglas et al. 2015), and they are continuously improving *space food* quality. The Italian company Argotec and NASA have established and designed over the years the Space Food Lab and the Space Food Research Lab to develop and prepare food, respectively, for future European and American astronauts. Systems/packaging engineers, scientists, biologists, nutritionists and dieticians merge their knowledge together in conducting investigations, experiments and plan projects for the Space Food products development.

In the first human space flights, food was usually freeze-dried, inserted into small squeezable tubes or developed as bite-sized cubes coated with gelatine in order to reduce possible crumbling. Astronauts have always had a wide choice for what concerns their meals, but it was only with the Apollo programme that food quality and varieties have begun to grow considerably. In the NASA website, there is an area dedicated to *Food for Space Flight* in which Apollo astronauts were first to have hot water, which made rehydrating foods easier and improved the food's taste. They were also the first to use the spoon bowl, a plastic container that could be opened and its contents eaten with a spoon.[2] At the same time, a small galley oven was used to heat food.

Original design criteria for the Apollo food, which were evolved from experience in food systems development for military, civilian and aerospace (Mercury and Gemini) use, are summarized briefly in the *Apollo Experience Report—Food System*, edited by NASA Technical Note D-7720 in 1974, in which there are some notes of *Original Food and Packaging Design Consideration* for food preparation (weight, volume, crew acceptance, preparation time, food-residue stabilization,

---

[2]https://www.nasa.gov/audience/forstudents/postsecondary/features/F_Food_for_Space_Flight.html.

Fig. 2.9 The astronaut Samantha Cristoforetti preparing her "bonus food" recipes in microgravity: whole red rice with peas and turmeric chicken (This picture has been taken by the video "Cooking in Space: Whole Red Rice and Turmeric Chicken" by ESA https://www.youtube.com/watch?v=4exaXdPKS3Y). Credits by NASA

wholesomeness, gastrointestinal compatibility, nutrient content, stability, physical, chemical, microbiological, contingency, shape and size), packaging (protection, identification, manufacture, weight, volume, function, material) and safety (through systematic procedures including the strict selection and control of raw materials, personnel, processing techniques, equipment, storage and manufacturing environments, and final assembly for spacecraft stowage). These criteria were modified to reflect experience gained with each Apollo mission and to meet updated and expanded mission objectives.

Food consumption in Space completely changed with the ISS development, where thermostabilized, refrigerated, natural form, irradiated, frozen and intermediate moisture meals are mainly consumed. Today, astronaut's foods are very similar—and in some cases the same—to those eaten on Earth with more than one hundred dishes to choose divided into three main daily deals. With so much variety available, it's hard to imagine choosing meals or being limited when embarking on a 6-month mission on the International Space Station (ISS). The Human Health and Performance Directorate's (HH&P) Space Food Systems Capability at NASA[3] is responsible for evaluating, producing and packaging food for each mission and developing menus. Food flown on orbit is not only required to be of high quality, it must be appetizing even when an astronaut's taste buds have changed due to fluid shifts from weightlessness. It must also be safe, easy to prepare and eat in a confined space with limited options for heating. The Space Food Systems team is

[3]https://www.nasa.gov/content/space-food-systems.

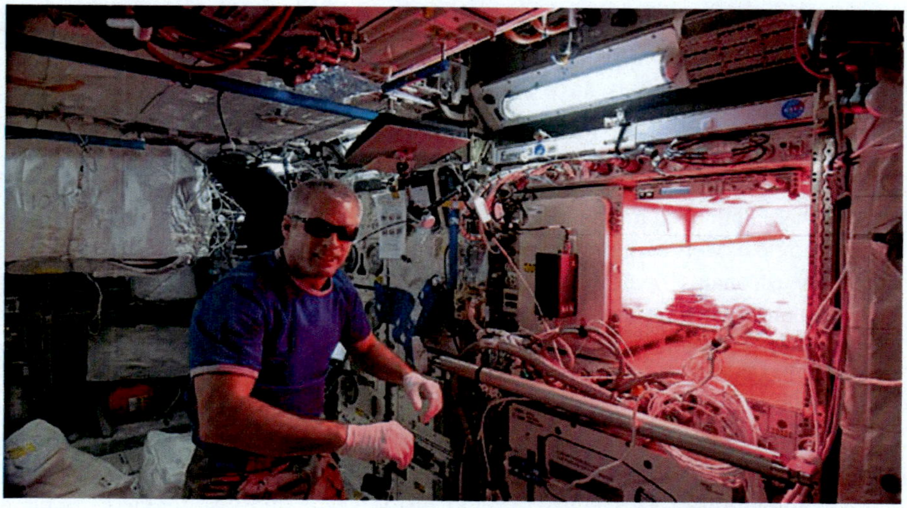

Fig. 2.10 The Veggie, Vegetable Production System, is a "space garden" on board the ISS for both emotional and practical purposes. Credits by NASA

required to meet the nutritional needs of each crew member while adhering to the requirements of limited storage space, limited preparation options and the difficulties of eating without gravity.

Food is such an important part of our lives and culture and plays a critical social and psychological role during the astronaut's stay on the ISS. As in the last years, ISS space meals are prepared by renowned chefs according to astronaut's needs and desires, for long space missions—for which it is not possible to receive supplies from Earth—it is necessary to think and plan how to directly grow vegetables and fruits that give them the nutrients needed. Recently, the Italian company Argotec produced the "bonus food" for the Italian astronauts Samantha Cristoforetti, *Futura* Mission in 2015 (Fig. 2.9) and Luca Parmitano, *Beyond* Mission in 2019.

There are some elements to consider that make critical the conservation of the food in Space. First of all, we have to consider that in Space there are two challenges to overcome: there is no refrigerator or freezer, we have to keep food at room temperature in conditions of high pressure and weightlessness. The last element to consider is the need to be able to keep food ready for at least 18–24 months, sometimes 3 years. Even packaging has some rules because we cannot afford large clutter or weights. For this reason, it is sometimes preferred the freeze-dried product to be stored in a box of 30 × 20 cm where we can keep stocks for long periods. There are also critical aspects regarding the confined environment: when the astronauts open the packages, they must not escape volatile parts and neither must fly crumbs or pieces of food that could end up in delicate instruments or even be inhaled. When using cutlery, the food must not "lose pieces" in the surrounding environment. In addition, no liquids can be transported. Orange juice, for example, is lyophilised and then hydrated before use.

From 2014, astronauts have the possibility to work with the so-called Veggie, Vegetable Production System, a particular "space garden" on the ISS to bring plants for both aesthetic and practical reasons (Fig. 2.10). We already know from our pioneering astronauts that fresh flowers and gardens on the ISS create a beautiful atmosphere and "take a little piece of Earth with us on our journeys" is good for our psychological well-being on Earth and in Space. They also will be critical for keeping astronauts healthy on long-duration missions.[4] Veggie is a simple and cheap plant growth chamber made by a flat-panel light bank with three different LEDs (red, blue and green) that enable the plant growth. It was created by the Orbital Technologies Corporation (ORBITEC) through a Small Business Innovation Research Program with the aim to provide a new resource for astronauts and researchers developing the capabilities of growing fresh produce and other large plants on the ISS.[5]

The evolution of astronaut's meals over the years, and the Veggie example are used to stress the importance in understanding which will be the next important steps and the following directions needed to establish a future base on the Moon. As already mentioned, food and nutrition are extremely necessary not only for the astronaut's physical well-being, but also for what concerns their mental and psychological status. Food is such an important part of our lives and culture.

Inside the Igluna 2020 initiative, described as a case study in Chapter 4 of this book, several groups are working together and developing new innovative and feasible ways to directly growth food on the Moon by adopting circular-economy systems needed to avoid waste in Space: from the biological conversion of human urine for food and bio-based oxygen production to the creation of fully automated systems for vegetables' growth and harvest both inside and outside the main lunar base. But once collected, how will these foods be transformed into meals? Which tools are needed to bridge the gap between food cultivation and consumption phases? As designers, we tried to approach these questions from uncommon points of view trying to consider differently the term food itself. We created a link between practical and psychological necessities in order to complete as much as possible the Maslow's hierarchy of needs.

Sleeping, for example, is solved by an equipment similar to an ordinary sleeping bag, or a berth in a cabin (Fig. 2.11). The astronaut must climb into it, tie it up tightly, in order to contrast the *sensation of falling*—due to the fact that the whole muscle system in microgravity is in a state of permanent relaxation as it does not have to support the body—and to avoid floating away in the air currents, once again because of weightlessness. It's also important to secure arms and legs to avoid them hovering as sleeping. To complete the equipment, there is also an eye mask and

[4]https://www.nasa.gov/content/growing-plants-in-space.

[5]https://www.nasa.gov/mission_pages/station/research/news/veggie.

earplugs to compensate for the lack of the privacy that is required when resting and sleeping and because on board there is a high level of noise due to the maintenance equipment of the ISS (Rockoff et al. 1985).

Most astronauts choose to sleep during Space Shuttle missions normally strap themselves into seats or attach sleeping bags to the floor, the walls or the ceilings. In the microgravity environment, there is no such thing as *up* or *down* which means it's just as easy to sleep vertically as you would horizontally back home. They usually avoid the cockpit since light from the sun can cause the area to heat up significantly and make slumber uncomfortable (Evans 2012).

Spacecrafts like the International Space Station (ISS) have pressurized cabins and are filled with the same kind of air we breathe on Earth, so the atmosphere on board is made to feel as close to sea level as possible. But microgravity causes astronauts to experience the effects of weightlessness and setting up a mattress on the floor can't be part of the plan because not only would the astronaut float away after dozing off, the mattress itself would also drift off, creating the potential for *midair collisions*. Because of this effect, astronauts could, theoretically, sleep almost anywhere in a spacecraft. Most of the crew on the ISS choose to sleep in their own cabin or in an ISS module. The American crew members' sleeping quarters are well ventilated to prevent breathing in the carbon dioxide you just breathed out, soundproof private cabin-for-one setups where an astronaut can have a little privacy and rest time, but much more could be done to increase the comfort and wellness (Mead et al. 1985).

Taking care of the personal hygiene is another largely unsolved problem. It is enough to consider that at present on board the International Space Station the possibility of washing or having a shower is not envisaged: cleaning up is managed by wiping with tissues dampened with detergent. The situation is worsened by the limited amount of clothing available for each astronaut, including underclothes, which does not allow for frequent changes, as well as the necessity of carrying out physical exercises for at least two hours a day—in order to contrast the loss of muscle mass and tone and the bone decalcification as we have seen in the previous pages—which increases the need to change clothes daily, because of extensive perspiration during aerobic activity. These two examples highlight the fact that not only are user expectations of comfort and well-being in Space not given due consideration, but above all that the primary necessities of every single human relative to strictly physiological needs, like sleeping and taking care of the personal hygiene, are neglected.

The only and unique "shower" designed for the Skylab, and not more in use, has not received much success from astronauts (Fig. 2.12). It worked in orbit, but, to the surprise of the author, many of the astronauts did not feel it was required nor needed. Because of the time to operate and clean up, few of the crewmen thought it worth the effort. The astronaut Owen Garriot reported: "The average time for a shower was about 1 h. You can clean up with a wash towel in 10 min or less. To me, I would skip the whole thing". Future systems should be simpler to operate. Here is a short description of the shower. This is an enclosure which uses continuous airflow as a gravity substitute to move the water over the crewman. A 6-lb

Fig. 2.11 Most of the crew on the ISS choose to sleep in their own cabin or in an ISS module. Credits by NASA

capacity water bottle is filled from the waste management compartment water heater, pressurized with nitrogen and attached to the grid ceiling at the shower location. The nitrogen gas pressurant expels water from the bottle through a transfer hose and a crew—operated hand-held spray nozzle. A soap dispenser provides the crewman with 8 mL of liquid soap for each shower. During the shower, this dispenser fastens to the ceiling with Velcro. A suction head removes water from the crewman and the shower interior. The suction head is connected by hoses to the centrifugal separator, which deposits the waste's water into a collection bag. A blower pulls the air from the separator through a hydrophobic filter that protects the blower. Another astronaut, Jack Lousma, commented in an official document recording the results of a project which sought to learn from Skylab experience to help the ISS design: "You had to mix water in a 3-quart container so much hot, so much cold. You had soap that was probably better used in a veterinary practice, because it made you itch. We sprayed water on ourselves with a sprayer and then had to vacuum it off with a suction device. One thing worth noting is that we were in this low-pressure environment and so whenever you got that water on you and went to dry yourself off, it got extremely cold because it was evaporating so rapidly. It took a lot of vacuuming to get all of the water out from inside of cylinder. You had to use a lot of towels to get dry. During our mission we usually just took sponge baths every night with a washcloth and a towel". Their comments are summed up in the NASA Technical Memorandum X-73073 *Skylab Lessons Learned as*

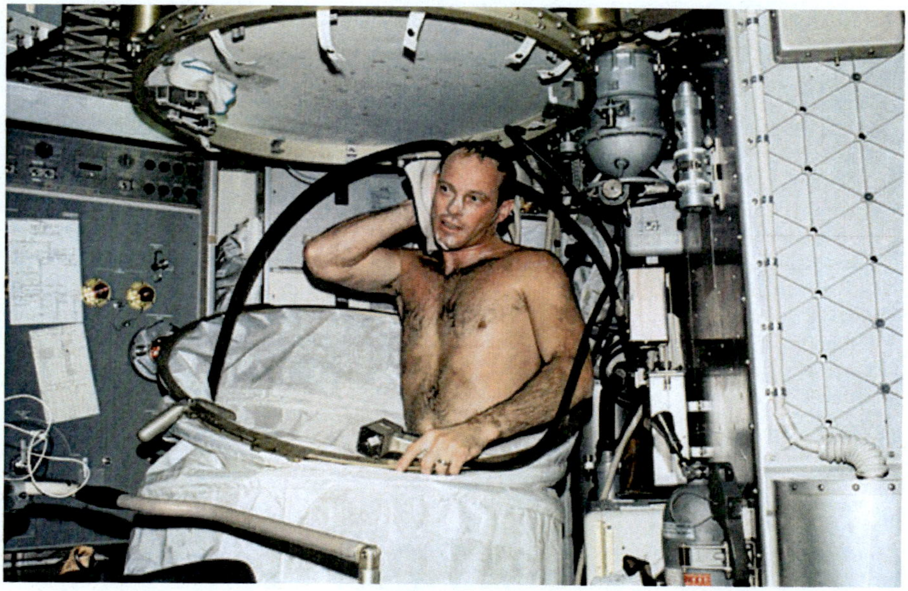

Fig. 2.12 The shower designed for the Skylab, and not more in use, has not received much success from astronauts for a lot of complications in use. Credits by NASA

Applicable to a Large Space Station produced by the Office of Space Flight Headquarter in Washington, D.C. in April 1976 (Johnson 1974).

These few examples underline that on Earth gestures, like taking a shower, shaving, as well as washing or cutting hairs, have to be completely re-invented in Space, together with new tools specific for microgravity conditions and confined environments. This is the task of Design for Space.

References

Antonutto G, di Prampero PE (1992) Human physiology in microgravity: an overview. Aerosp Med Biol 12(2): 145–147 (NASA SP (Series), Washington, D.C.)

Clark P (1988) The Soviet manned space program. An illustrated history of the men, the missions, and the spacecraft. Orion Books, London

Dominoni A (2002) Industrial design for space. Silvana Editoriale, Milan

Dominoni A (2015) For designers with their head beyond the clouds. Politecnica. Architettura, Ingegneria, Scienze, Maggioli Editore, Milan

Douglas G, Glaus-Late K, Kloeris V (2015) In: Brief J (ed) Space food systems

Evans B (2012) At home in space: the late seventies into the eighties. Springer

Gazenko OG, Shumakov VI, Kakurin LI, Katcov VE, Chestukhin VV, Nikolayenko EM, Gvozdev SV, Rumyantsev VV, Vasilyev VK (1982) Effects of various countermeasures against the adverse effects of weightlessness on central circulation in the healthy man. Aviat Space Environ Med 53(6): 523–530

Johnson CC (1974) Skylab experiment M487, habitability/crew quarters. In: Paper 133 of the Skylab results: proceedings, vol 1, 20th American astronautical society meeting, Los Angeles, 20–22 Aug 1974

Malerba F (1993) La vetta. Tormena Editore, Genoa

Mead GH, Peercy RI, Raasch RF (1985) Space station crew safety. Alternatives study—final report, vol V—space station safety plan. NASA Contractor Report 3858. Contract NAS1-17242, June 1985

Moore D, Bie P, Oser H (1996) Biological and medical research in space. In: An overview of life sciences research in microgravity. Springer, Berlin, Heidelberg

Peercy RI, Raasch RF, Rockoff LA (1985a) Space station crew safety. Alternatives study, vol I—final summary report. NASA Contractor Report 3854. Contract NAS1-17242, June 1985

Peercy RI, Raasch RF, Rockoff LA (1985b) Space station crew safety. Alternatives study—final report, vol IV—appendices. NASA Contractor Report 3857. Contract NAS1-17242, June 1985

Raasch RF, Peercy RI, Rockoff LA (1985) Space station crew safety. Alternatives study, vol II—threat development. NASA Contractor Report 3855. Contract NAS1-17242, June 1985

Rockoff LA, Raasch RF, Peercy RI (1985) Space station crew safety. Alternatives study—final report, vol III—safety impact of human factors. NASA Contractor Report 3856. Contract NAS1-17242, June 1985

Souza K, Hogan R, Ballard R (1995) Life into space. Space life sciences experiments. NASA Ames Research Center 1965–1990

Souza K, Etheridge G, Callahan PX (1991–1998) Life into space. Space life sciences experiments. Ames Research Center, Kennedy Space Center 1991–1998

Stuster J (1986) Space station habitability. Recommendations based on a systematic comparative analysis of analogous conditions. NASA Contractor Report 3943. Scientific and Technical Information Branch

Taylor TC, Spencer JS, Rocha CJ, Khan E, Cliffton E, Carr C (1987) Space station architectural elements model study. NASA Contractor Report 4027. Purchase Order A211774C, Jan 1987

Wichman HA, Donaldson SI (1996) Remote ergonomic research in space: spacelab findings and a proposal. Aviat Space Environ Med 67(2) (Alexandria, VA)

Chapter 3
Research and Design for Space Life

3.1 The Birth of a New Discipline in Space Design

To affirm with certainty and determination that Design for Space plays a fundamental and strategic role in human space exploration, it took me years of researches and projects that started from my Ph.D. in *Industrial Design for Space* at the Politecnico di Milano, after taking a degree in Architecture, at the same university, and becoming an architect.

When I started my Ph.D. researches in 1998—asking myself if an architect could have a role in *Designing for Space*, and which one—the Space field was the domain of scientists and engineers. Nobody was conscious of the importance of *well-being* to design proper environments and tools to facilitate life in Space and increase the comfort of human beings. No one even understood what Design was. Architecture was already something more familiar, even if in "Space language" architecture means *the global plan of an engineering system*, the structure of a whole Space program, and not the design of spaceships with expertise in habitability. But it was much worse for the Design field.

"Design" in English means "Project", and despite the amount of literature on the topic that has tried to clarify, affirm and analyse the *sense* of the word "Design", still today it is difficult to explain that Design is *something more than working on the surface of an object*; it's not just designing something generically, it's a discipline that embraces other disciplines, because the Design has a *multidisciplinary* nature.

The Design makes real ideas and gives shape to our world; it is the interpreter of the needs of people. Indeed, many times it is able to anticipate desires and to determine through objects and environments great changes in social behaviours and beliefs. Also, Design has a strategic approach. To this great misunderstanding about "Design", I should have added the word "Space".

A. Dominoni, *Design of Supporting Systems for Life in Outer Space*, Research for Development, https://doi.org/10.1007/978-3-030-60942-9_3

Fig. 3.1 The International Space Station (ISS) at the final configuration stage. Credits by NASA

What "Space Design" could mean? How I could have explained my intent? Space Design could have become a discipline? The beginning was not very encouraging. But I decided that I would try.

3.1.1 A Space Architect and Designer Meets Engineers

I looked for interlocutors in the aerospace world, space agencies and industries, engineers and scientists to understand better how to face the *hardware* of the architecture, the habitation modules, the equipment and tools for living and working in Space; physicians and astronauts who could help me get close to the *software* concerning the human being and the effects suffered in confinement environment and in microgravity conditions. My aim was to begin to speak their languages and make them understand what contribution I could give being an architect, a designer and a researcher who had just started a Ph.D. on a new discipline in Space Design.

From the scientific literature concerning the Space field I found, the only discipline close to design was *ergonomics*, but applied with an *engineering approach*, which means considering the human being "something that must work": astronauts had to adapt their selves to the machine, not vice versa.

Besides, astronauts have a lot of needs that machines do not have and that creates discontinuity and disruption in the development of the various activities and operations they have to perform on board: they have to sleep, they have to eat, they

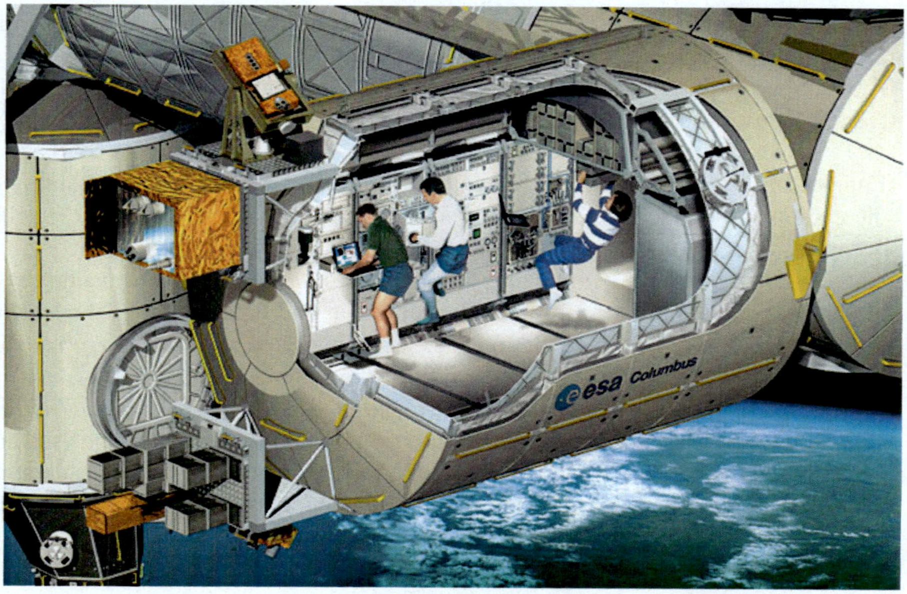

Fig. 3.2 The European Columbus Laboratory of the International Space Station (ISS) was realized by Thales Alenia Space for the European Space Agency (ESA). Credits by ESA

must take care of their hygiene and physiological needs… robots are by far preferred to humans according to the engineering thought, although on several occasions it is still proven that humans can find creative solutions, such as the Apollo 13 mission[1] and make the difference.

But I wanted to be positive and I thought that the coincidence of starting my Ph. D. in the same year in which the construction of the International Space Station (ISS) began might be considered a very exciting and promising point of start. In fact, it brought me luck and I was immediately rewarded for my efforts.

First, I realized that in Turin, just a hundred kilometres from Milan, Thales Alenia Space (in 1997 it was called Alenia Space) was designing and building over

[1]On 13 April 1970, Apollo 13 commander Jim Lovell told NASA headquarters: "Houston, we've had a problem" and began to explain the technical details of the failure. There was an explosion on board, and the ship was rapidly losing oxygen. I will begin for the crew and the whole of America, the odyssey of the difficult return to Earth, which should have been place 4 days later. All that remained was to abandon the objective of the Moon. The NASA control center in Houston decided for a passage around the Moon to push the Space Shuttle and make it take a route to Earth with a return trajectory "free", i.e. driven manually. In the lunar module, the astronauts lived critical hours that the whole world followed with bated breath: three men for 4 days in a space designed to accommodate two people for 2 days. The energy reserves were scarce, and even the water, the heat decreased. The carbon dioxide filters were not enough for three people, but by making do with what they found on board and with the guidance of the control center, the astronauts built an adapter that they called "the mailbox". Jim Lovell and Jeffrey Kluger then wrote a book about the mission, Lost Moon, which inspired the Ron Howard Apollo 13 film.

40% of the habitable pressurized modules of the entire ISS (Fig. 3.1), thanks to its extraordinary experience and technological know-how. Just think that today Thales Alenia Space—a joint venture between Thales (67%) and Leonardo (33%)—is the second industrial contractor of the orbital complex, after the American Boeing.

I contacted them immediately, and I got in touch with the Human Factors Engineering Department that it dealt with the aspects of habitability and usability of the equipment for the ISS. They were all engineers, apart from an architect and a designer, who, in order to *survive* and to integrate themselves, had adapted to the language of ergonomics, eliminating *artistic ambitiousness* like, for example, the importance of *soft interventions* as the use of colour and light to change the level of perceptual comfort of the environments, as well as the *emotional aspects* that impact directly on the performance quality of the astronauts' activities.

The *attention of the engineers was focussed on the human factors, and not on the human needs*, treating the astronauts as a sum of parties that can be analysed and decomposed in singular pieces, like those of a car, without taking into consideration the whole *person*, that is something more than many body parts put together that have to work in some way. The holistic approach was completely absent.

3.1.2 The Ability to Predict Use and Gestures

When I arrived at Alenia Space, they were building the European Columbus Laboratory (Fig. 3.2) for the ISS, and they were testing in the pool a dimensional *mock-up*[2] to make experiments of the usability of the tools on the outside of the habitation module that covered various Extra-Vehicular Activities (EVA). I tried to learn their language to drive the concept that *if the astronauts can live better, in a comfortable environment, considering in addition to the human factors also those*

[2]The *mock-up* is a model with the same dimensions and weight as the real module, but it is not made of the same materials. There are four stages in the development of the mock-up, even if usually all four are never constructed for one single project.

The first model represents the *shell*, the environment and overall volume, and it permits verification of its dimensions, movements, illumination and colours.

The second, more approximate than the first because it is made of inexpensive materials, is more precise in terms of the *internal configuration* of the structure: complex details are verified and possible movement between the components is foreseen, and it can be tested underwater. This model is often substituted by an electronic mock-up for testing volumetric relationships, assembly and disassembly of the components, etc.

Real parts are added to the third mock-up, which is very similar to the second one: these are parts that actually work from a *mechanical point of view*, such as connectors, switches, lights, energy and interfaces. However, they are made of inexpensive materials.

The last mock-up has real equipment that is useful for *training the crew* such as the possibility of carrying out test procedures of the various instruments that the astronauts will find on board during the mission but there is no redundancy, there are no doubles of instruments foreseen in the real configuration. Furthermore, the functionality of the instruments on the fourth mock-up imitates the real situation, but not of the quality of the instruments in terms of safety and reliability.

related to the psychological and emotional sphere, their performances and the results of the whole mission increase naturally.

Besides, I started to introduce a new methodology that I called *Use and Gesture Design (UGD)* which put in a systemic view the human being and the movements, the interactions and the gestures related to the equipment to use during scientific and maintenance activities, or experiments, and the environments. I wanted to join to the design process the analysis of use and gestures while imagining a new product that will be used in microgravity and in confined spaces, an experience that for all of us is unknown and it is not part of our daily life.[3]

Designing for Space was becoming for me something more defined, and as I affirm also today, it means *starting anew, applying a different logic for a different environment*, conceiving new instruments for uses and activities that Earth dwellers have difficulty in envisaging, but which on the whole presuppose a different relationship between our bodies, the objects and the surrounding space. Designing for confined spaces and for microgravity conditions require the *ability to predict use and gestures* that is to say foreseeing how an object will be used and in which ways, to be able to imagine, for example, the actions and the movements of the crew in relationship to the new Neutral Body Posture (NBP) that is assumed in microgravity, and foresee the physiological, perceptive, ergonomic, psychological and motorial requirements that will arise to the astronauts in conditions that are completely new and unknown to the human being (Dominoni 2002).

The most important and most interesting difference between *Designing for Space* and *Designing for Earth* can therefore be identified in a new design process in which it is introduced the capacity of use forecast, to visualizing gestures and movements that interact with equipment and tools in extra-terrestrial environments which cannot be compared to already known human experiences that can be amply verified. Defining the boundaries of the *UGD* methodology for Space helped me to expand and apply these principles also on Earth as a *spin-off of behaviours*, concerning the use and gestures of the crew on board, as well as of technologies, concerning the habitation modules and the whole architecture systems and sub-systems in Space.[4]

---

[3]The *Use and Gesture Design (UGD)* methodology is explained in the first chapter of this book and represents an innovation in approaching the development of the project, both on Earth then in Space, looking at the behaviours and the movements of the human being while interacting with tools and environments.

[4]If usually the *spin-off* term is used to describe a *technological transfer* from different fields of application, I started from the beginning of my experience using spin-off (from Space to Earth) and *spin-in* (from Earth to Space) to define *behaviours transfers*, because Design discipline concerns, before all, the human being.

3.1.3 Turning Microgravity into an Advantage

A case study that could help the readers to better clarify this new design process that introduces the methodology of *Use and Gesture Design (UGD)* along with the ability to predict the uses of the object in the contest for which it is created—and that I had the opportunity to create and experiment concretely, coming into contact with Alenia Space in 1998—concerns the development of a system of tool-carrying containers, which Alenia Space asked me to design, and which I have called *Portable Caddy System* to be used as a toolset to support the routine maintenance of the European Columbus Laboratory of the International Space Station (ISS).

The project was part of a broader programme based on the development of concepts and Design solutions in order to increase the efficiency of systems to support movement, habitability and work in confined environments and in conditions of microgravity. As part of this research programme, the design of a tool holder system for routine maintenance of the ISS Columbus European Laboratory is located in the area of required equipment "Outfitting Complements" or prosthetic objects whose main purpose is to increase the efficiency of the crew's activities.

Compared to the existing containers used at that time on the ISS that are as rigid toolboxes sized to be integrated into the racks[5] room dedicated to storage, I studied a system of soft containers, made of fabric, that was portable, and also easily wearable by astronauts. The project follows the principle of *turning microgravity into an advantage*, not a limit, trying to exploit the new potential that can have weightless objects floating in space.

The idea was to imagine the containers close to the astronauts, as useful assistants, but independent, of the kind of *console* that could become rigid—if you had to lean and fix the tools that were not needed at that time, avoiding that they begin to float in the air, or that they were attached in a scattered way to the body of the astronauts with the Velcro®—and return soft, to be rolled, once used the tools and stored them in the *Portable Caddy System*, to occupy the minimum of space possible. My inspiration was the soft containers of fabric that are used to store jewellery, or those for cutlery, which are very similar in shape and use. I thought that instead of jewellery or cutlery, you could also imagine storing utensils, of different sizes for different uses, in the same way. Apart from the lightness and the saving of space, the project allowed—thanks to a flexible structure inside the fabric made of harmonic steel, to create the necessary rigidity, once unrolled the soft container, to take the form of a rigid console, a surface close to the astronaut, to support and fix the various tools, a sort of flying working plan.

The system included several pieces: in addition to the console, the largest piece, I had also proposed smaller pieces, wearables, for example, around the thigh, useful when astronauts were in confined and limited spaces that they could not use the

[5]The *rack* is a sort of modular cabinet which covers the walls of the habitable modules of the ISS. Its structure is conceived to host experiments, equipment and tools useful for the whole functioning and maintenance of the International Space Station (ISS).

console, or even if they need to use only a few pieces for certain operations, and not the entire collection of tools. This was my first project in which to experiment with the intuitions of the new methodology of *Use and Gesture Design (UGD)*—that is, the analysis of the movements more suitable for the use of the soft toolboxes and the tools punctuated by precise gestural sequences—and an extraordinary opportunity to increase my predictive use and gesture ability—to simulate the movements of astronauts before on Earth, through a preventive verification of the possible uses and gestures in relation to the body, objects, the activities to be carried out and the Space environment—that has been articulated intertwining with the parallel development of morphologic, functional and performance characteristics of the *Portable Caddy System*.

The project for the *Portable Caddy System* can be traced back to the area of research conducted by the Human Factors Engineering Department of Alenia Space on the human performance in conditions of microgravity. Like the most projects of equipment and objects supporting astronauts' activities, they were taking on an important role, although in Alenia Space they were reduced to an ergonomic analysis by applying pre-constituted parameters and tables, without considering the diversity of approach and behaviour between one astronaut and another, and without applying innovative solutions: their principle was to proceed for small steps, with incremental innovations, without thinking of new projects or new uses.

For this reason, I believe that my *Portable Caddy System*—created through a new Design approach, which introduced the importance of combining the design of the objects with that of its possible uses in an environment without weight—has been welcomed with great interest by the Human Factors Engineering Department.

3.1.4 Design Research Methodologies for Space

The *Portable Caddy System* allows me to introduce instruments and methodologies that belong to the Design discipline, as the *Use and Gesture Design (UGD)* methodology that I created specifically for Space Design, *hybridized* with the engineering approach of *Task Analysis* used by the Human Factors Engineering Department of Alenia Space.

The starting point of this Design Research coincided with a predominantly *empirical survey* to consider the experiences previously acquired in this field of intervention evaluating the weaknesses and strengths of the system currently in use. The goal was to *re-think* a system of containers designed specifically for Space, avoiding adapting elements in use on Earth with the limits that derive from this setting. This approach required a Design Research methodology based on an *inseparable connection between theory and practice*, where practice means the ability to simulate results through the prior *verification of uses and gestures in relation to objects and the environment*. In the meantime, I involved Usag, an Italian company specialized in producing tools and utensils, to support me to develop the project and realized the prototypes of the *Portable Caddy System*.

The following are the main points of investigation that I defined together with Alenia Space in order to improve the quality and the efficiency of the already used equipment for the same functions:

1. Analysis of the movements and the human activities carried out within the habitable volumes, applying the *Use and Gesture Design (UGD)* methodology, and provided for the ordinary maintenance of the module to get to define the design requirements of the *Portable Caddy System*.
2. Study of available spaces and existing anchoring systems to facilitate movements and human activities for routine maintenance carried out within the habitable volumes, with particular reference to those planned for the International Space Station and the European Columbus Laboratory.
3. Analysis and identification of the advantages and disadvantages encountered by comparing the different anchoring systems in use.
4. Analysis and identification of areas requiring anchorages for the *Portable Caddy System* containers and the tools contained.
5. Study of the system currently in use and analysis of the advantages and disadvantages encountered on board the International Space Station.
6. Identification of the design requirements indicated by NASA as required (Ref. NASA-STD-3000 and ESA-RQ-013): number of tools and their type, number of procedures provided with listed in sequence the tools needed for each operation, number of postures required for the various routine maintenance operations and their type, suitable materials for the environment, functionality, comfort and freedom of movement.
7. Development of a new system of portable and wearable tool containers for routine maintenance called *Portable Caddy System* and based on the innovative concept of taking advantage of the peculiarities of the environment and microgravity conditions, considering until now the greatest limit of living and working in Space.
8. Realization of the first prototypes by Usag on the basis of the design brief.
9. Verification of requirements and design assumptions through testing of dry and water *Portable Caddy System* prototypes.
10. Realization of subsequent prototypes by Usag based on test results.
11. Verification of requirements and design hypotheses through experimentation and tests with parabolic flights.

Analysing the state of the art I focussed on the system currently in use, a *Tool Box*, developed by Alenia Space and DMR for the Italian Space Agency (ASI) and tested during two missions on the Russian Orbiting Space Station MIR: Euromir '94 and Euromir '95. *Tool Box* was designed to support crew members in the study and development of the "On-Orbit Testing of Crew Support Equipment Study" (Musso et al. 1996). It consists of a 2-mm-thick opaque anodized aluminium container to which Velcro® strips and elastic belts have been applied on the outer surface to provide temporary attachment of the tools and other necessary objects. The inside of the container is intended for tools. These were selected taking into account the offer on the market, and the consequent adaptations of the individual parts to the conditions required by the space environment were reduced to a minimum. The list of tools provided for the container is generic: the selection was made considering a generic use of the container for routine maintenance operations and for emergency situations. Each tool has a ring attached to which is tied a rope that is fixed by Velcro® to the belt or wrist of the operator to prevent the tools from getting out of hand and start to navigate in Space. The total weight of the fully equipped tool container is 2.7 kg. The overall analysis to evaluate the efficiency and validity of the *Tool Box* has given rise to some considerations that have served to detect the major limits of this system:

- poor visibility of tools;
- poor solution of fixing tools inside the container;
- poor solution of attaching tools to the operator because the ropes tangle around the body limiting movements;
- high weight.

The state of the art included also a document NASA—Boeing which collected various equipment for astronauts including Equipment Bag Assembly (EBA): it was a container consisting of a large expandable central space with small external and internal pockets and an outer band to carry it on the shoulder. The fabric was made of a semi-transparent mesh held at the corners by hinges and Velcro®.

My Design Research would try to overcome the limits found in the *Tool Box* system by transforming them into project objectives-requirements. The innovation points of my *Portable Caddy System* were given by the objectives I put as results of the state-of-the-art's survey:

1. The main objective identified was to provide permanent storage of tools in the racks of the International Space Station (ISS) and at the same time to provide temporary support to tools and equipment used for ordinary activities on board maintenance. This was the first innovation: *to provide a system of containers able to perform the two functions, permanent storage and temporary use of tools*. Until now, in fact, systems have been designed separately for the permanent storage of equipment in racks (Ref. Boeing *Soft Stowage* Human Space Logistic System, Catalogue Issue 1) and portable containers for the temporary use of the same equipment. A system of containers that performs both functions results in a saving in volume, weight and a consequent reduction in equipment.
2. A further step forward has been the study of container supports that allow to be transported with ease and positioned in different ways and angles adaptable to the different areas, often poorly accessible, of the ISS. This is the second innovation: *to provide a system of portable and wearable containers able to guarantee flexibility in use and immediate availability of tools*.
3. The third objective was to try to solve and overcome the limitations of the system currently in use by making the tools visible from the outside of the containers. This is the third innovation: *to provide a system of semi-transparent containers able to make tools visible* (Fig. 3.3).

Fig. 3.3 Two prototypes of the *Portable Caddy System* designed by the author make visible the tools increasing efficiency and usability of astronauts in Space. Credits by the author

4. The fourth objective has been to try to solve and overcome the limits of the system currently in use (Ref. APM *IVA Tool List Maintenance*, Technical Note, Columbus Team, Alenia Space Division, 29.09.99) allowing the operator to carry out the various activities without losing the tools or being limited by the fixing system of the same. These are the fourth innovation: *to provide a system of containers able to ensure an effective attachment of the same tools to the support and the operator*.

These four innovations gave rise to other more specific objectives that contributed to increase the comfort, the efficiency and the usability of the *Portable Caddy System* in Space:

- the reduction of the volume and specific weight of containers compared to those currently in use;
- flexibility to adapt to different spaces and maintenance operations;
- easy access to tools by the operator;
- the wearability of containers to facilitate their transport in orbit during various activities (Fig. 3.4);
- the softness of the material to fit the body and the different spaces available.

Considering the set of objectives and requirements defined above, this Design Research has made it possible to define a system of portable main containers

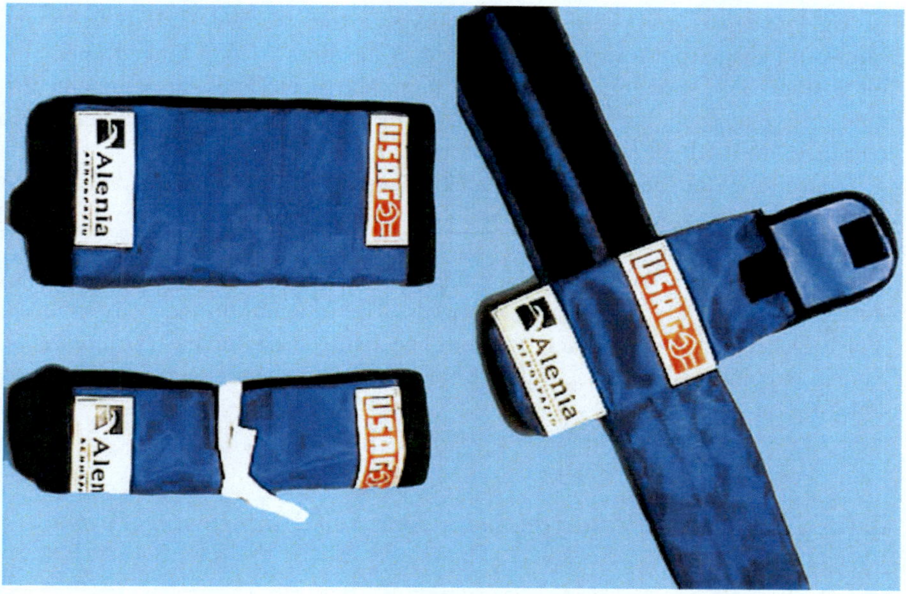

Fig. 3.4 Two prototypes of wearable containers of the *Portable Caddy System* designed by the author facilitate tools transport during operations on ISS. Credits by the author

rollable and closable which, once opened, can be completely stretched, or only partially, flat and in every direction, thanks to two bands of harmonic steel with return of shape inserted in the fabric. The harmonic steel bands have the task of stiffening the support and transforming it from a portable container to a work surface during use: this solution allows the container to be fixed to the internal structure of the module and at the same time to adapt to the limited and sometimes difficult to reach spaces of the intervention areas inside the European Columbus Laboratory (Ref. Boeing, *Flight Crew Support and Integration (FCS&I)*, Section of the *Space Station Operations Data Book (SSODB)*, December 1997).

The *main containers* have inner pockets semi-transparent made with a net to divide the various tools and ensure the total visibility and accessibility of the tools during use. The internal pockets have flexible and standardized dimensions to accommodate different tools. The accessibility of the tools through the pockets is resolved with elastic bands that facilitate the extraction of the tools but at the same time prevent their escape. There are also external pockets to place small accessories. The attachment of the container system to the internal structure of the module is through Velcro® belts that roll up by self.

The *portable and wearable containers* (Ref. NASA-STD-3000) have a semi-rigid shape, pre-formed and rounded at the sides, with a hinged opening on three sides. The stiffening system with harmonic steel bands, also present in the main containers, allows the wearable container, once opened, to become a small console with the tools well visible and close at hand.

Concerning the operator-object relationship, the portable and wearable containers have been studied and designed to be positioned adherent to the thigh, at the waist and transversally (shoulder-waist), compatible with the type of operation to be carried out. The possibility of placing them on the arm and on the leg was ruled out due to poor visibility of access to the tools. The fastening to the body is through Velcro® belts that roll up self-sealing. The fabric used for the *Portable Caddy System* is composed of Betacloth and Nomex fibres. There is also an antibacterial finishing on the tissue.

3.1.5 To Verify Prototypes with Dry and Underwater Tests

Alenia Space was interested in developing study models and prototypes of the Portable Caddy System to verify the usability (at that time the term *usability* was not widespread and they used *ergonomy*) and the effectiveness of the new system of fabric containers, soft and wearable. As usual in developing projects and as already explained before, I decided to involve a *terrestrial* company, specialized in the production of tools, Usag, for the realization of prototypes of the soft bags.

I led first *dry tests*, so-called in *Space language* to describe the tests that can be done on Earth and that are useful for small objects or equipment, as a rack interface, in which the interaction with the human being does not change a lot when in microgravity. Dry tests are the last procedure that is carried out on the ground before performing tests underwater or in parabolic flights is referred to as a *dry test*. It is very useful for defining in detail the operations to be carried out, such as screwing and unscrewing components and structural parts, in order to verify that everything is functioning correctly and to avoid jeopardizing the verification possibilities offered by the other tests.

Then, I had the chance to led *underwater tests* in the big swimming pool of Alenia Space called the Neutral Buoyancy Facility (NBF) the best place on Earth for simulating weightlessness, in which a mock-up of the European Columbus Laboratory has been built, to test astronauts operation during Extra-Vehicular Activities (EVA). There are several peculiarities that are characteristic of tests carried out underwater which are important to point out and that concern the relation between the body, the gestures and the movements and a good playground to apply the *Use and Gesture Design (UGD)* methodology:

- if the equilibration of the subject is appropriate in relation to the depth, the simulation of the absence of gravity is very realistic;
- the friction of the water influences rapid movement in a significant manner; contrary to *parabolic flights*,[6] gravity is not completely absent, so all the objects that are not balanced "fall" to the bottom of the swimming pool if they are not appropriately tied, instead of "floating away" as happens in orbit;
- the duration of the tests can be lengthy, with no limitation on the simulation procedures, which instead are strictly limited in the case of parabolic flights;
- there are no problems of "airsickness" typical of parabolic flights, which limit the productivity of the subjects and therefore the percentage of the success of the tests.

There are some limits, however, regarding the configuration and the design of the simulators to be introduced into the swimming pool: it is not possible to use electric apparatuses, and therefore the operation of the flight system or of electric tools cannot be simulated; metal structures must be able to resist corrosion and allow adequate water flow, in order to minimize the hydrodynamic inertia; materials must be waterproof and may not vary in performance once immersed in water.

Apart from the possibility to experiment the microgravity in the same way, the simulation of the Space environment in water is in fact rather approximate: there is no real sensation of microgravity; the subjects involved in the tests essentially cannot communicate very well with each other; the lighting level is very different from the real one and it can be difficult to see well when performing activities; the conditions of movement are altered by the tendency—inherent in the subjects—to "swim".

This kind of underwater simulation is, however, particularly useful for the tests involving the crew activities and equipment characterized by significant volumes as well as complex sequences of operations, for instance: large apparatuses to be transported, assembled and removed, assembled, long and complex procedures, to be evaluated in an integrated manner; activities requiring the involvement of several subjects; Extra-Vehicular Activities (EVA), or when it is useful to have a mock-up of several elements, or sections of elements, of the International Space Station (ISS); accessibility to equipment whose location cannot be simulated in parabolic flights, also for reasons of safety.

Coming back to my project, after this digression on the validation systems useful for testing new Space products, the prototypes of the *Portable Caddy System* have had several opportunities to be used and verified during the experiments as support to the planned activities of Alenia Space. The first results commented by the test

---

[6]The *parabolic flights* are currently the simulation system that most closely imitates the effective conditions of microgravity experienced in orbit, but for very short intervals of time, approximately 30 s of microgravity each parabola with an Airbus A300. The bigger limit is the need to subdivide the operative procedures into many separate sequences of operations lasting no more than the period of one parabola. A standard flight campaign lasts 3 days, with 30 to 40 parabolas carried out on each day, considering that between two successive parabolas they pass from 2 to about 4 min.

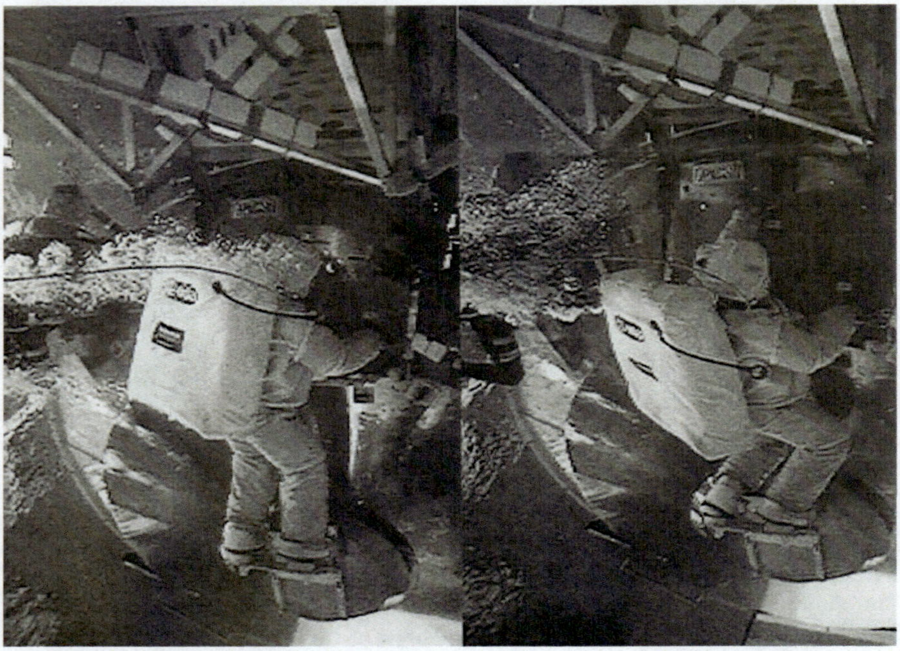

Fig. 3.5 I took these photos during the tests of comfort, efficiency and usability of the *Portable Caddy System* conducted underwater inside the Neutral Buoyancy Facility (NBF) of Alenia Space in Turin. Credits by the author

subjects provided interesting input for further development of the *Portable Caddy System* according to its future use in conditions of microgravity.

The tests conducted in the laboratories of Alenia Spazio, *dry* and *underwater* (Figs. 3.5 and 3.6), have demonstrated the effectiveness of the system during use and, in particular, have been extremely functional, in both portable main containers and wearable containers, the following activities related to the *Use and Gesture Design (UGD)* methodology:

1. portable main containers:
   opening and unrolling, securing the *Portable Caddy System* inside the module (even if in water, in horizontal position, the stiffening system with harmonic steel bands meets too much resistance and is therefore not reliable), closing and rolling, the fastening of the *Portable Caddy System* inside the module by Velcro® (rather than with elastic clamp), the repositioning of the tools in the *Portable Caddy System*;
2. portable and wearable containers:
   the opening, the adhesion to the leg, the accessibility to the tools, the repositioning of the tools in the wearable *Portable Caddy System*, the closing.

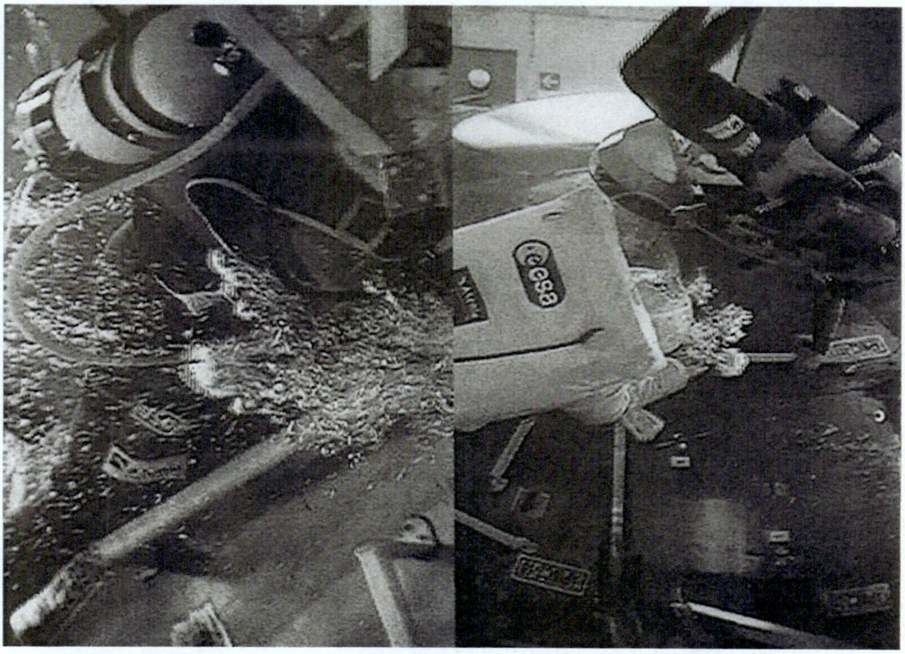

Fig. 3.6 In these images, it is possible to see the portable and wearable containers on the thighs of the astronauts while simulating ordinary maintenance of the ISS during both Extra-Vehicular Activities (EVA) and Intra-Vehicular Activities (IVA). Credits by the author

In both portable main and wearable containers (1. and 2.), the choice to use soft material such as fabric instead of metal or plastic was successful for the extreme ease of opening and closing and for the best adaptability to body shapes when it has to be worn (Fig. 3.6).

3.1.6 Co-design with Astronauts

It was a great new experience for me to have the opportunity to receive feedback from astronauts testing my *Portable Caddy Tools System*, even though I did not receive, as I thought, useful guidance for improving or modifying prototypes.

The *briefings* and *de-briefings* I led with them[7] provided for me to create very specific questionnaires that left no room for dialogue. The open questions I had

[7]The *briefing* and the *de-briefing* are activities that involve astronauts to check the products during the design process, before the realization of the prototypes (briefing activities) and after to have conducted tests of usability (de-briefing activities) in order to give opinions and suggestions to improve the performative qualities of the project.

prepared didn't give me answers that would help me better understand their needs, and in this particular case, to understand if my system was working. I had to go back to my ability to imagine *how it could have been if* I had lived in Space, without gravity. Later, I would find that, with a few exceptions, astronauts are trained to perform very complex and detailed procedures that do not require their participation in the design process. Today, we would have called it a *co-design* process. For them, creative skills are not contemplated, although I still believe that designing for and together with our users, especially astronauts who experience what it's like to live and work in Space, is important, if not indispensable.

We will see later in this same chapter how the first Space architects and designers, from the design of the Soyuz to the Skylab up to the International Space Station, had a different feedback from the crew, certainly more participation and enthusiasm from cosmonauts than astronauts, that perhaps for training and provenance are closer to a military approach that is based on orders that must be executed quickly and that often do not include an active and proactive contribution.

3.1.7 Disruptive Feasibility Studies with Design Companies

For the moment, we continue with my experience that in the same year 1998 is enriched with other opportunities for Space Design: the Italian Space Agency (ASI) launched the first *Call for the Technological Utilization of the Space Station* (which will be unfortunately also the last until today). I participated with two proposals to develop feasibility studies of new products to be used by astronauts in Space in which I had involved for the technological part, some Italian industrial partners considered leaders in their respective fields, and above all that they had a strong vocation to do research and innovate their products, and for the *spatialization*[8] of the prototypes Alenia Space. The objective of the ASI was to select the most innovative feasibility studies that would become, through standard procedures, experiments to be tested on board the International Space Station (ISS). At the end of the selection's procedure, I was very happy to discover that both my proposals had been accepted and financed. It was the first time that in the field of Space were involved *terrestrial* companies, and overall *design-oriented*, to participate to a Space programme that would have required the realization of prototypes to be tested on board the ISS, and in ASI they were very curious to see the results involving private companies that did not belong to the Space field like all the others who did not belong to the space sector like all the others who usually participated in the calls.

[8]We utilize to call in Space jargon *spatialization* the sum of all the procedures referred to NASA Technical Standard 3000, specific for human space flights, and tests that make a product *ready to be launched in Space* respecting requirements of human factors, habitability and environmental health.

The first proposal, in which I involved Benetton Group—the Italian company very famous to have launched during the 1980s the sweaters round neck with lots of colour gradations—was a feasibility study that I called *VEST. Integrated Clothing System for Intra-Vehicular Activities (IVA)* is used by the astronauts on board the International Space Station (ISS). The project was made of different typologies of clothing focused from one hand, on the characteristics of comfort and wearability of the garments, and from the other hand, on new smart textiles researches, in particular related to the thermoregulating and antibacterial properties of the fibres, to increase the well-being of the crew during human space flights (Dominoni 2005). Also, the project foresaw the integration of a system of sensors into the garments to monitor the biomedical parameters without being invasive for the astronauts and require their active participation. The project *VEST* will be treated as a case study in this book in chapter 4 *Space Design Between Research, Project and Education*.

The second proposal, in which I involved TechnoGym—a worldwide Italian company leader in designing fitness machines and defining wellness scenarios—was a feasibility study that I called *AGILE, Attrezzi GInnici LEggeri for Intra-Vehicular Activities (IVA)* to be used by the astronauts on board the International Space Station (ISS). The project was made of a system of fitness machines, more light and friendly of the existing ones at that moment (a VELO cyclo ergometer and a trade mill) and aimed to increase the efficiency of the physical exercise, reduce the time of training, and also make easier and pleasant the two mandatory hours per day dedicated to maintain muscle mass and bone structure. *AGILE* implied many studies and researches on physiology specific for microgravity environment in which appropriate physical exercises become for astronauts one of the most efficacy countermeasures to the loss of muscle mass and bone decalcification that astronauts encounter living in Space.

3.1.8 Space Design is Multidisciplinary

For me, of course, starting to devote myself to projects that had an almost immediate effective response was a great joy and the beginning of a series of new and exciting activities, in which as project leader and Principal Investigator I found myself interacting with experts in the various disciplinary fields, who had different backgrounds and languages, and who often had difficulty understanding each other. For the designer, accustomed to dealing with traditional production realities, and above all terrestrial ones, the Design Research in the Space field is only possible with the presence and constant participation of agencies and industries specialized in the Space sector that are able to provide you with all the necessary information to address the project assignments from time to time.

I soon discovered that a fundamental feature of the Design discipline is its multidisciplinary nature, the ability to speak different languages, to relate to the scientific community of reference, to collect data and requirements to translate them into solutions and to act as a bridge between the project and the human sciences, using the group as an accelerator to the creative process. This topic will be exploited later on in this same chapter in the paragraph *Designer's Perception at ESA*.

Taking into account that the Design embraces the entire world of artifacts, I would have found myself soon designing, after the experience of the toolboxes, *Portable Caddy System*, a clothing support system, *VEST*, a fitness system equipment and machines, *AGILE*, as well as a washing machines that it is assumed to be able to wash without water, systems for the preparation, the consumption and the conservation of the food, systems for the illumination and the personal hygiene, and all items and tools regarding the whole habitability in the Space.

3.1.9 Crossing Space Knowledge Through Design

The identification of the new areas of research presupposed for me, from the lens of Design, scrupulous analysis of the Space environment and the specificities that most affected everyday life, as the activities the crew carried out, or they could have performed in the future, on board the International Space Station (ISS).

Along with the scientific literature, unfortunately very poor as regards the contribution of Design discipline in the development of real projects that could be tested by astronauts, I was looking for information on ongoing research programmes that involved human space missions to identify strategic development lines that could have been crucial in the design process, with the aim to find possible areas of convergence with Design. I was trying to complete the puzzle on the state of the art of a sector, the Space, still too small and young to be able to count on reliable statistics conducted on the astronauts' behaviours and experiences in confined environments and in microgravity conditions.

The experience gathered by the soviet and American human space missions in half a century of activity had not produced enough data to be formalized and translated into shared scientific knowledge; indeed, I found *vague information, and sometimes even contradictory, because of the subjectivity of every astronaut*. For example, I was struck by the fact that in Space, sight could become less acute—30% of astronauts, even after short-term missions, reported a decrease in visual capacity—but above all that some reported seeing even the faintest colours, in *pastel tones*. The causes of this phenomenon are not yet entirely clear but, according to the physicians and medicals, it could be triggered by the absence of gravity that makes the body fluids concentrate in the lower limbs and the head, thus reducing the pressure of the retina. So, *which colours should you choose for microgravity environments?* Very bright shades to be perceived as "normal" compared to earthly experience? Or soft colours to favour calm and relaxation? How much does colour affect the environment of the ISS to support the activities of astronauts, to promote concentration as well as sleep? No study so far has been tested aboard the International Space Station on the power of light and colour to facilitate the performance of astronauts on board.

The *lack of reliability of the experience acquired and the little knowledge of the Space environment with its effects on man* was due to the small number of human space missions and astronauts involved—the Space conquest is still in a pioneering phase—and, in part, to the duration of the same missions which, except for rare occasions of some continuous months of stay on board, apart from the Russian cosmonauts (Gadzenko et al. 1982), in most cases did not exceed 2 weeks. I'm talking about the early 2000s when the scientific experiments to be conducted on the International Space Station (ISS) could count on a 7/8 day of human space missions called *Taxi Flight* for the brevity of the period. Today, the astronauts' missions last approximately 6 months, and in some cases have even exceeded a year. The scientific researches have also made progresses and now we know more about the effects of the Space environment on the human being (Wichman and Donaldson 1996).

The most comprehensive study of long-term microgravity permanence remains the one conducted on Scott Kelly's body during and after the *One Year Mission* between 2015 and 2016 (one consecutive year in orbit on ISS) and that of his twin remained on Earth. If, on the one hand, the data reveal unexpected details about the adaptability of our organism, on the other hand, there are still many unknowns and it is difficult to generalize. Much of the information that emerges may relate to the specific way Scott Kelly's body reacted, and it may not be true for all astronauts, nor for the general population. It is difficult to say if the results can be extended to the next human space missions to the Moon and Mars, well beyond the 400 km distance from the ISS home, where you can enjoy all the protective effects of the Earth's magnetic field.

For the designer, the knowledge of the Space environment and its effects on the human being is essential to be able to identify the requirements to generate ideas that can offer the crew the actual benefits in terms of comfort and efficiency in the performance of the activities on board.

The exploration of a new area such as Space, in which I could enter with the role of researcher and designer, was complex and innovative both from the technological than the sociological point of view. This implied the presence of people who had devoted their lives to study and training, hoping to be chosen, at least once, to participate in human space missions that would have taken them beyond Earth, out of our planet, and to face risks still little known to the scientific community. I am not thinking only of the most extreme experiences, in which the entire crew has lost their lives, but I would like to recall that even today very little is known about the harmful effects of cosmic radiation and cell degeneration in the absence of gravity, especially in the medium and long terms (Moore et al. 1996).

I also discovered that *Designing for Space* required the industrial designer to commit, just as important as designing, which concerned the *strategic organization of the new scenarios* that were on the horizon (Dominoni 2002). And I was increasingly convinced that Space would have a decisive influence in the near future on people's lives and behaviours, enhancing health as wellness. I needed to dedicate an equally important commitment to that of designing which concerned the organization of new scenarios and the management of contacts necessary to relate to the relevant scientific community. The identification of new areas of research requires a thorough analysis of the Space environment, its peculiarities and the consequent influences on life and all the various activities that the astronauts carry out and could carry out on board, not to mention the ability to imagine spin-offs through new applications on Earth.

The first action I considered essential was to integrate to the scientific literature—at that time the only book available that could be close to my interests was *Living Aloft* (Connors et al. 1985)—and papers, and information on the state of the art of the field with the research programmes in progress trying to maintain a direct contact with the space agencies and identify the strategic lines of development of the human space flight's exploration that could have given me decisive suggestions to Design for Space, and perhaps even inspirations on what might have been the possible areas of convergence with the Industrial Design discipline. It is not enough to have many new ideas if they cannot be tested directly with astronauts. It is essential to be known and to make clear what are the potentials of Design in the Space field.

In order to be able to work as a researcher and designer of Space equipment and environment, it is therefore essential for the designer that the scientific and industrial Space community is constantly informed about nature and the role of Design, to know the potential and possible applications of this discipline in confined environments and microgravity conditions. In this regard, a path that I would have immediately pursued is to participate in symposia, calls for competitions and workshops, organized by national and international space agencies, to disseminate the results achieved and research developments, and progressively gain credibility and interest in Space scientific community, consisting of agencies, research and development centres, product certification and validation bodies, and Space industries. Very often, the correct identification of the interlocutor can determine the quality of the project results (Dominoni 2015).

At the International Workshop *ESA ExploSpace. Space Exploration and Resources Exploitation*, organized by European Space Agency (ESA) and led from 20 to 22 of October 1998 in Sardinia, I presented my first paper[9] *IVA Clothing Support System* concerning the first results of a *Clothing Support System* designed for the crew of the International Space Station (ISS), able to improve the comfort of living and working on board through the wearability of garments, aesthetics, thermal stability and body hygiene (Dominoni 1998). I was the only woman presenting a scientific paper and the only designer among many engineers and physicists among which Carlo Rubbia (Nobel Prize in Physics 1984) just before me. My presentation was noted with curiousness and interest and for me was a fantastic occasion to meet a lot of extraordinary people and start to build my network of academic and scientist coming from Space field.

[9]The Paper *IVA Clothing Support System* received the *Scientific Award for Young Researchers* by the European Space Agency (ESA): In appreciation of Annalisa Dominoni scientific value presenting the paper "IVA Clothing Support System" at ESA ExploSpace Workshop, Space Exploration and Resources Exploitation, Cagliari, Sardinia.

After that lucky episode other projects began, such as the *Feasibility Study of Habitability on board the ISS* commissioned by the Italian Space Agency (ASI) that asked me to imagine a new scenario with equipment and tools thought specifically for the Space environment, including various activities as the preparation, consumption and conservation of food, the body care and personal hygiene, the entertainment activities during the free time on board, the physical exercises, the possibility to grow plants in microgravity together with living, working and sleeping main activities on ISS. It was a big work of designing and forecasting use and gestures in a confined environment and microgravity conditions which led me to gain more and more experience in Space Design.

Then, the more exciting work for Space was to become Principal Investigator of two experiments on board the ISS, *VEST, Clothing Support System for the International Space Station*, during the *Marco Polo Mission* in April 2002, and *GOAL, Garments for Orbital Activities in weightLessness*, during the *Eneide Mission* in April 2005, both performed by the astronaut Roberto Vittori. I do not want here to describe more the experience because it is presented as the case study in chapter 4 of this book *Space Design Between Research, Project and Education.* After these two experiments, other projects of research and design have followed each other but now I would like to stop here with my personal experience that gave me the opportunity to introduce approaches, principles and methodologies of the new discipline in Space Design, and expand my gaze beyond time and history to retrace some milestones of Space Design.

3.2 Space Design Was Born by a Woman

Few know that was a woman who had the value to give shape to the human space flights for the first time in the world, an architect, a designer who took care of well-being of the cosmonauts introducing in her projects' harmony, beauty, colour and semiotic languages balancing with human needs, functions and technology. During the *Space Race* in the 1950s and 1960s, the task of designing the look of the soviet union's booster rockets and orbital laboratories fell to one woman: Galina Balashova. For the budding architect in the midst of a militarized rush into Space, the work was also a chance to bring the principles of architecture and design into places they had never been before.

This woman defined the style of the whole soviet space program, from its logos to its satellites. Her delicate and rich watercolour and pencil illustrations of habitation modules, interiors, furniture, control panels, ergonomic studies and the vast paraphernalia which embraced everything from badges and logos to stationery and satellites are testaments to functionality, and have served multiple generations of Russian spacecrafts up to the present: from the design for the Soyuz capsule to the Salyut and the Mir space stations, including the Buran programme and the soviet

Fig. 3.7 A watercolour realized by Galina Balashova who designed habitation modules, interiors and furniture for the Soyuz capsule and the Salyut and the Mir space stations. It this drawing the typography design for the name placement on the outer shell of the Mir. Credits by Galina Balashova Archive

reusable spacecraft. But they are also gorgeous works of art, like relics from a nearly forgotten future. Her work, much of it once top secret, has itself nearly been forgotten, even within Russia.

Galina Balashova, born in 1931, began her career as an architect by stripping off the decorations from Stalinist buildings that had fallen foul of the Khrushchev reforms. The classical language of socialist realism was firmly out by the mid-1950s when she started, but the designs still had to get built. She finished her career in the early 1990s as the pre-eminent architect of human space exploration. Balashova began working for OKB-1, the soviet space agency, designing residential buildings for the scientists and engineers working on the soviet space program. As the only architect in the team, she was then poached to work on the interiors of the manned spacecraft that came in the wake of Yuri Gagarin's first brief space flight in 1961. Recently, in 2015, an exhibition at Deutsches Architekturmuseum (DAM) in Frankfurt/Main named *Design for the Soviet Space Program. The Architect Galina Balashova* enlightened her great work and underline the importance of Design in Space (Meuser 2015). What is surprising is the sheer range of her architect's work. From interiors and Space furniture to the graphics on the sides of spaceships (Fig. 3.7) and on the inevitable soviet space programme medals, her hand is everywhere. Her impeccable watercolour renders completely defined the image of the soviet space program, both in Space and on Earth, and on stationery. She used

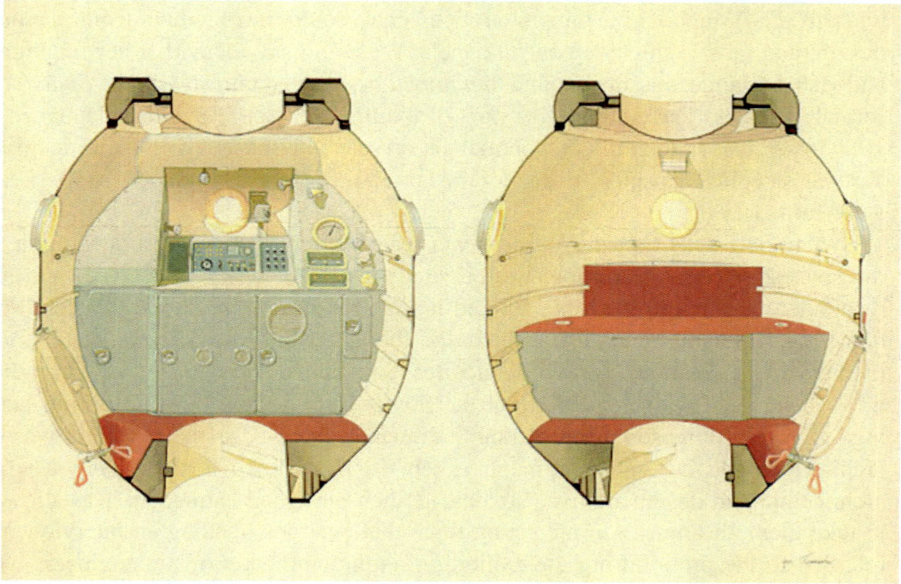

Fig. 3.8 She introduced semiotic and colour languages to meet the cosmonauts' needs and make sure they felt at home in the small capsules with *minimal design* that could be sweet, functional and easy to use. Illustration Study of the Soyuz Living Area. Credits by Galina Balashova Archive

to say that *a good architecture always depends on the same rules, and it doesn't matter whether it's about a house or a spaceship*. And I would add, it doesn't matter whether it's about an object, or a tool, a garment or a fitness machine.

As architect and designer, her projects give shape to the whole Space Nathan Rogers *From the Spoon to a City* in which the Milanese architect and critic intended to express the breadth of the territory to express the breadth of the territory project of our profession, which allows acting simultaneously on different scales of complexity concerning the human being.

3.2.1 The Language of Semiotic and Colour in Space

There are some striking insights into Space Design for zero gravity. It was thought, for instance, that in zero-gravity conditions, cosmonauts wouldn't bother about orientation in the Spacecraft, but Galina Balashova realized it mattered and, in a *semiotic coding*, introduced *colours schemes* (Fig. 3.8) that ensured floors were always dark coloured and ceilings light, so cosmonauts could have reference points to orient themselves. Furniture was designed to look *domestic* rather than *Space Age*. While product designers were playing with aluminium and moulded plastics shaping a world inspired by Space scenario as sci-fi movies, she designed glass-fronted sideboards, integrated beds that could have come from nautical design, and radiogram-style control panels. We could say today that her functional and visual attitude was attributable to *minimal design* in contrast with the *Space Age* formal excesses. The complex interiors of the first Mir Space Station—launched in 1986 to serve as mankind's first permanent presence in Space—with their ingenious vertical sleeping capsules, were, for instance, meticulously planned by her. She went further.

When the Soviet Space Program team realized that there was no decoration on the spaceship walls, Balashova was invited to paint a series of pictures and she responded by painting the remembered landscapes of her childhood, watercolours depicting Russian countryside, which she thought would be a reminder of home memory to the cosmonauts. "These pictures" she related "no longer exist, since they were incinerated with those parts of the Soyuz Capsule that burn up during landing".[10] The only reason her remarkable drawings survived is that, as there was no architecture and design department, except for Balashova herself, there was no architecture and design archive. So, despite their classified nature, she was allowed to take them all home. An interesting topic that emerges, reading an interview she released to the press during the exhibition, is the confusion of the engineers who manage the human space missions in distinguishing the role of *Art* from that of *Design*, or *Architecture*: One day, Konstantin Petrovich Feoktistov asked me to do a sketch for the interior of the *living module* for a spaceship. Actually, Feoktistov first asked artist Viktor Petrovich Dyumin, because he thought that this was a task for artists. Dyumin pointed out that designing interiors was something architects did. And so, one weekend I found myself working away drawing a sketch.[11]

The *living module* that Galina Balashova faced when she started her engagement was composed of a drawn with two red boxes in the plans. She wanted to *look at the cosmonauts needs to make sure they felt at home* in the small capsules creating a harmonious small room around them that could be *sweet, functional and easy to use*. The room needed to be outfitted just as if it were an apartment: with a bed, a table, chairs, a toilet and a shower.

In Space, we do not have any references regarding top and bottom, and it is very difficult to orientate. Balashova drew up two proposals. In the first one, the case in the Soyuz Capsules, she gave the space a *floor* and a *ceiling*. The floor was in green, the ceiling bright blue and the walls' sides bright yellow. In the other proposal, the case of a project she made for the Lunar Orbital Craft LOK, it was not to distinguish clearly between top and bottom, but to structure the Spacecraft by means of different volumes. The Soyuz Capsule variant won because it is easier to work in rooms that had a clearly defined top and bottom, and it is easier to train cosmonauts flew into Space in identical, true-to-scale mock-up. In some of her drawings, there are pictures on the spaceship walls, maybe pretty superfluous in space travel, where

[10]https://www.iconeye.com/opinion/review/item/12306-galina-balashova.

[11]https://www.stylepark.com/en/news/only-the-watercolors-burned-to-nothing.

every added gram and every extra centimetres of space counts, and overall, where functional and security aspects are predominant of all the other choice.

3.2.2 Balancing Human Needs and Functionality with Beauty

In Space, there is microgravity, and the human body reacts differently than on Earth, and there is a confined environment, but we can anyway to check the dimensions and the proportions of an interior of a spaceship through a mock-up analysing the relation of interface between human beings, available spaces and objects. As today we use to test with *analogues* the feasibility degree of a project, Galina Balashova as well faced with the interior of a submarine to have an idea of a confined environment in which people live for long time, without natural light and environmental stimuli, without privacy and the possibility to go out. “Another challenge was how to fasten documents and equipment in place: in zero gravity everything simply floats around otherwise. So, we clad the inside surfaces and all objects with Velcro® fasteners. The coloured wall cladding, the floor and the ceiling, it was all made of Velcro®. That way, you could attach objects simply anywhere, and just as easily detach them again. Only the chair and bed covers were made of normal fabric—the cosmonauts fastened themselves in place using belts. Their suits would have been damaged far too quickly under the permanent strain of Velcro® fasteners”.[12] The use of Velcro® is still currently the most efficient and the least invasive system useful to fix objects and tools to the interior surfaces of the spaceship, or to the bodies of the astronauts, mainly thanks to its simplicity and its minimum volume and weight. As well as considering functional aspects, she asserted that the most important skill for an architect or a designer is *a good taste*. That means “To find the right proportions in the balance of humans, space and architecture. And that you can always reconcile function and beauty—he’d cribbed the latter from Palladio” (Meuser 2015).

In this book, I drive the idea that *Design approach can add an aesthetic value to ergonomic and technology requirements*, and the projects of Galina Balashova are very good case studies that show how to *conciliate the world of engineering with the world of beauty*, which is so important for the human being’s harmony. Balashova was gratified by the cosmonauts who liked the interior design of the Soyuz and the Mir. Only once did one request brighter colours: he said it was a little too dark in the Space Station, because to save weight, only the bare minimum of lighting had been provided. So, she installed more lamps. Not only she was satisfied by the cosmonauts’ appreciation, but her success was transferred in orbit to the

[12]https://www.stylepark.com/en/news/only-the-watercolors-burned-to-nothing.

astronauts: "After the Apollo-Soyuz Program, our cosmonauts said the american astronauts really liked our interior design because it was so comfortable. *The american section was evidently far more tech and less homely*. That made me feel proud, as engineers don't think of the architecture only of the technical side to things. The best thing about my work was the freedom I had: No one told me I had to take this or that colour, do that this way, and whatever. I was able to take the decision I thought were the right ones as nobody was in the slightest interested in what I did".[13]

3.3 A Designer to Create the Skylab

When inside NASA, for the first time, the question of *habitability* was being asked designing the Skylab laboratory, the only one that would have items designed specifically for this unique space mission, Raymond Loewy, was already a famous designer and was chosen for this task. Loewy worked on functional styling for a variety of mass-produced products for 40 years, besides designing stores, shopping centres and office buildings. He gave shape to the bottle of Coca-Cola, the packaging of Lucky Strike, but also the Air Force One for President John F. Kennedy and the Greyhound bus, according to its streamline's philosophy and the ability to face the project in different scales.

Skylab was the United States' first space station[14] launched by NASA, a complex and complete *orbiting home* and *scientific laboratory*, an exciting adventure with all the drama that could be packed into its three manned flights, for about 24 weeks between May 1973 and February 1974. Major operations included an orbital workshop, a solar observatory, Earth observation and hundreds of experiments. It was conquest, of human-made hardware, a difficult and challenging environment, and even of ideas, and it was a severe test of man's capability to analyse, solve problems and make innovative repairs in a hostile and unforgiving environment. Skylab was ingenuity but also innovation. The programme was initiated with hardware developed for other programmes, modified for this Space odyssey, and supplemented by *items designed specifically for the conduct of its unique mission, many of them concerning habitability*, like the first and unique "shower" made to be used in Space.

---

[13]idem.

[14]https://www.nasa.gov/mission_pages/skylab/missions/skylab_manned.html.

3.3.1 From Functional to Comfortable Spacecraft

Habitability is, as one NASA designer remarked, "a nebulous term at best", one "not usually found in the engineer's vocabulary". Besides factors within the engineer's usual responsibilities, such as the composition and temperature of the atmosphere and the levels of light and noise, habitability also encompasses the ease of *keeping house*, the convenience of attending to *personal hygiene*, and the provision for *exercise* and off-duty *relaxation*. Experience and intuition both suggested that these factors would become more important as missions grew longer. Looking ahead to space stations, NASA designers needed basic information on these problems of living in Space (Johnson 1974).

Early Spacecraft had been designed to be operated, not lived in. Weight and volume limitation in the Mercury and Gemini Capsules—the epithet, though despised by crews, was apt—meant that only the bare requirements for protecting and sustaining life could be provided. Michael Collins, the pilot on Gemini 10, compared the two-man Gemini craft to the front seats of a Volkswagen. That tiny space was home for Frank Borman and James Lowell for 14 days on Gemini 7.

The Apollo command module, though just over twice the volume of Gemini, was still primarily a functional spacecraft. Some improvements made it a bit more pleasant—hot water, for example—and its extra space gave the crew of three astronauts some freedom to move around and exercise stiff muscles; but few concessions were made to mere comfort. For the most part, "astronauts accepted whatever discomforts were inherent in their spacecraft, unless they interfered with performances, what mattered was accomplishing the missions".

When early planners looked ahead to orbiting space stations, their attention was devoted to problems much more pressing than crew comfort. Of 41 papers presented at a Space Station Symposium in 1960, only one addressed the question of making the "space station a pleasant place to live". This paper (Payne 1960) noted that operating an orbiting station would be much like keeping a lighthouse and discussed some of the factors that would have to be improved so that people could be induced to go into Space. Some of these factors were intangible, said the author, but they were no less important for that. Nine years later, the situation had changed little. Spacecraft technology still occupied the engineers' attention, while the questions of everyday living were left for someone else to look after (Compton and Benson 1983).

Raymond Loewy was the first industrial designer called by NASA at Marshall Space Center (MSC)—which was involved in an experiment entitled *Habitability Crew Quarters* having the objective of obtaining design criteria for advanced spacecraft and long-term space stations—to give *some reasonable degree of creature comfort*. He produced a formal report in February 1968, *Habitability Study A. A. P. Program*, citing many faults in the existing layout and suggesting a number of improvements.

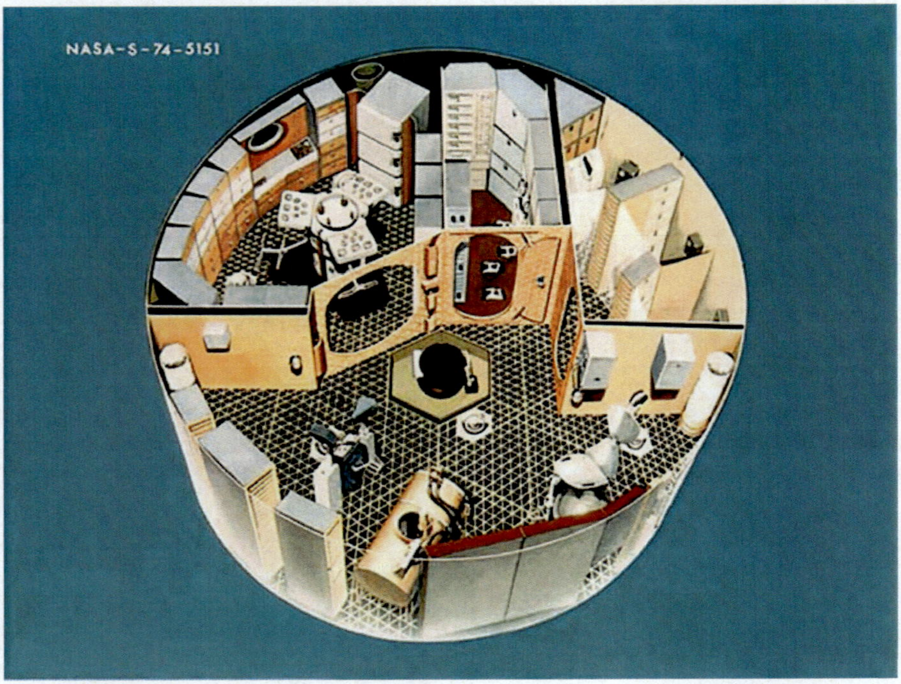

Fig. 3.9 Raymond Loewy designed a flexible layout for the *Habitability Crew Quarter* of the Skylab improving the colour organization and lighting to change the perception of space. Credits by Raymond Loewy Foundation

The interior of the Skylab laboratory was poorly planned, while Loewy thought that a working area should be simple, with enclosed and open *areas flowing smoothly as integrated elements against neutral backgrounds*.

While he found a certain honesty in the straightforward treatment of interior space, the overall impression was nonetheless forbidding. The basic cylindrical structure clashed with rectangular elements and with the harsh pattern of triangular grid work liberally spread throughout the laboratory. It served to separate the habitation module in three principal parts, and also for anchorage through special outer soles on astronauts' footwear. The visual environment was badly cluttered. Lights were scattered apparently at random over the ceiling, and colours were much too dark. This *depressing habitat could, however, be much improved simply by organized use of colour and illumination*. He recommended (Loewy 1981) a neutral background of pale yellow, with brighter accents for variety and for identifying crew aids, experiment equipment and personal kits. Lighting should be localized at work areas, and lights with a warmer spectral range substituted for the cold fluorescents used in the mock-up.

Fig. 3.10 Raymond Loewy persuaded NASA to design a large round *window* (on the top of the picture) in the wardroom of the Skylab to allow the crew enjoying the view of the Earth from Space enhancing recreational value. Credits by NASA

3.3.2 Improving Habitability and Recreational Values

Loewy proposed to create a wardroom, a space for eating, relaxing and handling routine office work, better yet, the floor plan should be made flexible by the use of movable panels, so that different arrangements could be tested (Fig. 3.9). These suggestions were received at MSC with a certain amount of perplexity (Belew 1977). And since none of the astronauts who had examined the mock-ups had attached any importance to such things. They cared less about styling and appearance than efficiency, and they wanted a Spacecraft in which they could do their jobs without a lot of petty annoyances. They were, in fact, somewhat disdainful of the attention given to such amenities as interior colour schemes.

In June 1968, a new principal investigator was appointed to work on habitability issues, Caldwell Johnson, Chief of Spacecraft Design in the Advanced Spacecraft Technology Division at MSC. He first looked at the Skylab laboratory convinced Johnson that habitability had been given no thought at all. In the course of their work, he and his colleagues had built up a store of information on design factors for all kinds of crew activities under circumstances of confinement and isolation, but their data might as well not have existed. So, he translated habitability from testing different design concepts to create an operational system that would reduce the chores of daily living to a level *entirely incidental* to human space flight operations. He proposed to deal with nine major components of habitability: environment, architecture, mobility and restraint, food and water, clothing, personal hygiene, housekeeping, communication within the spacecraft, and off-duty activity. By systematizing the man spacecraft relationship, Johnson hoped to bring some engineering rigour into an otherwise chaotic field (Johnson 1974).

As it happens in all the projects, at the end the final result is a compromise between the proposal of designers and the *resistance* of the commitment: some habitability recommendations of both Raymond Loewy and Caldwell Johnson were accepted, and had a great impact of the Skylab laboratory structure, like the rearrangement of the floor plan to provide a *wardroom*, and the addition of a large *window* (Fig. 3.10) to allow the crew to enjoy the view from orbit, because *its recreational value alone would be worth its cost on a long space mission.*

Also, the clothing system changed: Johnson had to persuade the crews that the test pilot's traditional *one-piece flight* coveralls were not suited to long-term living in the workshop. In this, they acquiesced, but they would not give up the pockets on the lower trouser leg, ideal for a pilot strapped into an airplane cockpit, but as Johnson believed, a useless impediment to moving around freely in zero-g. Johnson designed a *three-piece uniform* to which a matching *jacket* could be added; it was practical and attractive, but he was disappointed when the crews spoiled the effect by covering the shirt and jacket with name tags and badges.

When it came to matters of purely personal preference, such as off-duty relaxation and entertainment, Johnson was content to let the crew have their choice. He proposed an *entertainment centre* in the wardroom, equipped to show movies, or provide music, but it drew no enthusiastic response. Considering that the Skylab was the first space station in which human beings could experiment the condition of freedom to be weightlessness, *the main entertainment interest of the astronauts was living the microgravity and look at the Earth's magic beauty*.

Skylab's crewmen found that, within the roomy laboratory, and without the impediment of gravity's pull, they could perform feats that even the most talented acrobats on Earth could not duplicate. "Mobility around here is super—Conrad reported—nobody has any motion sickness. Every kid in the United States would have a blast up here". The cavernous workshop, 8.2 m long and 6.7 m in diameter, invited exploration. In Space, they soared around their home, embellishing their flights with flips, cartwheels and inventive gymnastic exercises (Belew 1977). So even if Skylab designers had provided recreational equipment to vary the routine of long hours in Space—there were dart sets, without sharp points, playing cards, balls, books, exercise equipment and a tape player—much of this went unused. Crew preferences tended strongly towards reading and recorded music—provided everyone could have his own private tape player; musical tastes were quite disparate (Compton and Benson 1983) and Earth gazing during orbital daylight and star watching during orbital night became principal diversions.

Thinking at the efforts of both designers of Skylab—Loewy and Johnson, respectively, outside and inside NASA, to create an environment more comfortable and beautiful to increase the well-being of astronauts—I can only feel close to their frustration that it is not just the feeling of to be considered superfluous and not indispensable, as engineers are. The disappointment is that some astronauts, fortunately not all, who are our users, were completely indifferent to Loewy and Johnson's efforts to improve their comfort. An opposite reaction is to the cosmonauts that appreciated all the design proposals of Galina Balashova to make the Space environment more harmonious and comfortable.

3.4 International Space Station: Design Versus Engineering

The project of the International Space Station (ISS) did not involve a designer or an architect from outside, as was the case for the Soyuz Capsules, the Salyut and Mir Space Stations, and the Skylab laboratory. The entire design of the habitable modules of the ISS was done within the engineering departments of space agencies, mainly NASA and ESA, in collaboration with the aerospace industry, of which the major contractors are Boeing and Thales Alenia Space.

The reasons for this change can be traced back to the fact that the ISS is an international project of the United States, Russia, Canada, Japan and the Space Agencies of 11 member countries of the European Community, in which each country has contributed, independently, to realize different *pieces* of the International Space Station, in a different time. The assembling of the ISS in orbit began in 1998, with the launch of the Russian cargo block Zarya—which in Russian means *dawn* and became one of the cornerstones of the ISS—and started from two central pressurized modules made by Russia and USA providing the ISS with energy, propulsion and space during the early stages of its assembly.

The ISS remains uninhabited for 2 years, and it is necessary to wait for the year 2000 to assume the form (or at least the functions) of a Space *house*. In July of that year, the launch of the Russian Zvezda module takes place, which hooks to Zarya and Unity. Zvezda integrates the rooms of the base which is now equipped with kitchen, toilets, dormitories, oxygen generators and carbon dioxide recycling plants, equipment for the training of astronauts and for the transmission of data. Everything is ready for the arrival of the first crew (Fig. 3.11).

Both Russia and the United States (US) space habitation modules are different in equipment layout and capabilities: launch, docking and modular systems. In the beginning, the US module was a sort of *plug-in* to the Russian module (without windows), which was self-sufficient about energy, limited resources, but sufficient. The US instead has a different concept: all the habitable modules need to be together to function as they are not autonomous spacecrafts. The European Columbus Laboratory, jointed in 2008, recalls the US internal layout that is organized per modular approach: bays, racks and drawers.

Fig. 3.11 From the bottom to the top: Unity, Zarya, Zvezda and the Progress M1-3 spacecraft seen from the Space Shuttle Atlantis in 2000. Credits by NASA

Looking at these first examples of configurations adopted by the international partners to build the ISS, it would have been difficult to imagine one commitment to a single designer to give shape and identity to the entire space station. But maybe, the right reason, and probably the main one, is the spread of a common thought that still considers Space still a pioneer sector, and in which the development of the habitation modules is the domain of engineering.

The designer still fails to emerge as a key figure necessary from the early stages of the project development, because the engineers do not consider priority aspects related to habitability, as we saw in the case study of the Skylab laboratory, in which although there was a part of NASA that was opening up to the concepts of comfort, the internal resistances to the various departments regarding the proposals of Raymond Loewy were still many, like wondering what a *window* was for, even if then, in the end, they realized. Even if the International Space Station today has the *Cupola*, developed and built in Italy by Thales Alenia Space, after more than 30 years from the Skylab, we are far from the regular involvement of Design competences in Space even if it is much bigger than the window of the Skylab. The *Cupola* (Fig. 3.12) is a spectacular robotic control room that transported to the International Space Station (ISS) with the STS-130 mission aboard the Endeavour, which was launched on 7 February 2010. The glass if 15 cm of thickness, no distortion or refraction, a *work of art and technology*. The *Cupola* lets astronauts see and work through seven windows, with a 360° view all around the ISS that allows to see the Earth, and also check the robotic arm. It has been hooked to the Node 3

Fig. 3.12 The *Cupola* of the International Space Station (ISS) can be considered a *work of art and technology*. Credits by NASA

Tranquillity, also developed by Thales Alenia Space, and is used as a quite spectacular Flight Control Center of the ISS during spacewalks, spacecraft manoeuvers or works that require the use of the robotic arm of the ISS. Everything is ruled from ground, there are no pilots on board and astronauts are scientist as in a laboratory.

The purpose of the present space programmes is to provide an environment where the space travellers of the present and the future can live and work efficiently and effectively forever longer continuous periods, also in anticipation of future interplanetary journeys, as a result, we are moving towards habitable systems that are self-sufficient and autonomous complements from land supplies. But, on the subject of habitability, however, the vision of Design is quite different from that of engineering, especially concerning the *meaning of habitability and wellness*.

3.4.1 Industrial Design at NASA

I had the chance recently to interview Marc Cohen who started working at NASA as architect before the beginning of the International Space Station Program and was able to give real contributions to increase the habitability in Space. His testimony confirms my thesis of the *dichotomy between Architecture and Design versus Engineering*. He was the second Space architect in NASA's Ames Research Center experimenting the opportunity to offer ideas, after Maynard Dalton at Johnson

Fig. 3.13 The *Suitport* EVA access facility mock-up during field tests, designed in the early phase of the ISS advanced development programme faces the important needs regarding the transition between inside and outside the pressurized habitation modules. Credits by NASA

Space Center (JSC), who started at the beginning of the Apollo programme around 1963. Here I report part of my interview to him completed during the month of June of this year 2020 that starts with his experience in NASA at the beginning of 1980s. "I always thought that any other professional architect in my job could make comparable contributions. When I think about that time in the 1980s and early 1990s, it was the most exciting time in my life, the perfect job for me in the best organization in the world at Ames. However, I become exhausted just remembering how hard I fought to implement my ideas into the program.

The International Space Station was really the last NASA space programme that was wide open at the beginning and allowed for real architectural design creativity. In contrast, unfortunately, the Orion allowed very little opportunity for architectural or human factors creativity besides the controls and displays because the NASA administrator at the time it started thought that replicating the Apollo command module was the best way to go. Orion has twice the pressurized volume and is rated for up to about 30 days in Space, with all new electronics and ECLSS are new, but otherwise it is much the same idea. The Artemis lunar Gateway has even less openness because NASA head quarter decided to build it with predetermined non-competitive contracts for existing hardware". Along his projects that were realized and flew in Space or achieved serious hardware prototyping, he drafted the original fabrication drawings for what became the first ISS equipment racks during a

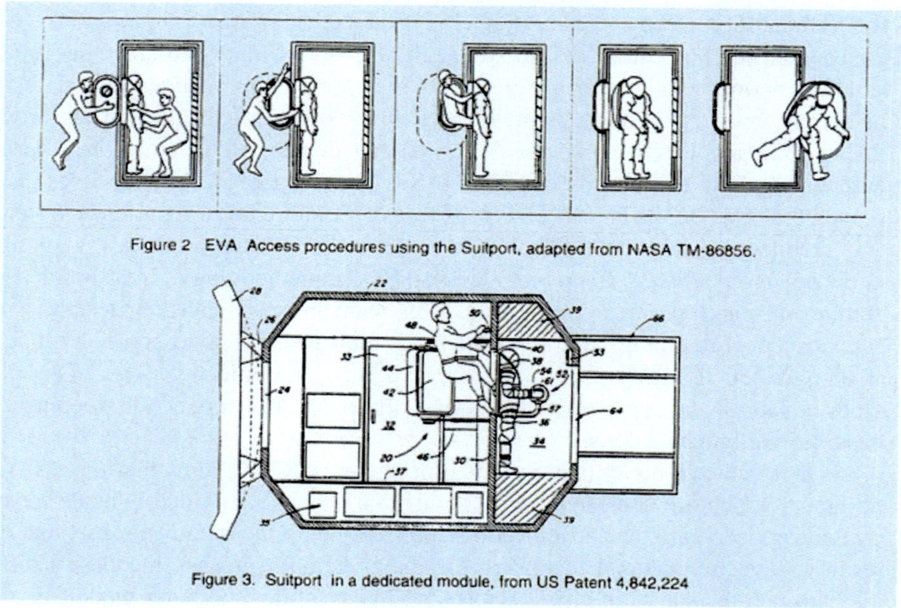

Fig. 3.14 Sequences of the predicted movements performed by the astronauts using the *Suitport* where we can see the procedures description comfort and the efficiency during the transition. Credits by M. Cohen, NASA

week at Marshall Space Flight Center (MSFC). The MSFC people built a whole module full of them for the US lab mock-up that became the Destiny lab module.

But the more his interesting project in which we can see the contribution of Design in Space field is the *Suitport* Extra-Vehicular Activity (EVA) access facility (Fig. 3.13) originated during the early phase of the International Space Station (ISS) advanced development program. It grew from a recognition that the construction and operation of Space Station Freedom (SSF) would require several thousand hours of EVA time. To make the best use of SSF crew time, it will become necessary to make the entire space suit donning and doffing, and airlock egress/ingress as safe, rapid and efficient as possible (Cohen 1995). The *Suitport* found its first practical use as a terrestrial application in the NASA Ames Hazmat vehicle for the clean-up of hazardous and toxic materials. In the Hazmat application, the *Suitport* offers substantial improvements over conventional hazard suits, by eliminating the necessity to decontaminate before doffing the suit. In the EVA access facility procedures using the *Suitport*, we can see an example of the *Use and Gesture Design (UGD)* methodology in which the transition from inside of the ISS and outside of the extreme environment is designed and predicted in all the possible movements performed by the astronauts (Fig. 3.14).

Confirming the trend of the International Space Station (ISS), in which industrial design's contribution of the habitable modules was an activity developed inside the engineering departments of space agencies in collaboration with the space industry, the Habitability Design Center (HDC) introduced at NASA around 2005 is a conceptual human-centred design studio. It was born initially to solve secondary small ergonomic issues, then progressively increased its involvement in the design of actual products destined to pressurized lunar rovers or the ISS itself (Taylor 2008) gathering together expertise in industrial design, architecture, and system engineering. The HDC[15] comes from NASA human space flight with a designapproach focussed on the *human factors* as a design tool to develop products,systems, and architecture for Space. The team is comprised of professionals withdirect knowledge on *Human Centered Design (HCD)* methodology, industrialdesign, architecture, and systems engineering. This team uses advanced Spacecraft concepts, rapid prototyping, and *mock-up* fabrication techniques to solve the unique challenges of living and working in extreme environments. The HDC provides advanced concepts for habitats and spacecraft interiors while keeping the needs of the human first and foremost.

At first view, in the online presentation of the HDC, it seems that the skills of designers which are considered most significant by NASA, which is made mostly by the engineers, are the capacities to *visualize* quickly ideas and concepts that can result very useful to attract customers: “Using an iterative process, the team takes an idea through the stages of artist sketches, to blueprinting, computer models, scaled prototypes and finally full-size *mock-up* fabrication. This process allows for stakeholders to provide feedback early in the design process and to gather critical usability assessments”.

Deepen the topic it appears clear that the HDC was an integral partner in the design of a vast number of products ranging from dining tables for the ISS to ergonomic seats for the Orion and Space Exploration Vehicle (SEV). Leveraging their expertise in human factors, HDC designers are often called upon to evaluate hardware to improve *usability*. Typical design questions include, can an astronaut easily reach all the critical controls if they are in casual clothing or fully suited? Does a seat conform to an astronaut regardless of their height and body size? Can a crew member anchor themselves in microgravity to easily eat at a galley table? Addressing these challenges early on in the design phase speeds, and often results, in a better product. The HDC team is often called upon to help oversee field tests of vehicle prototypes as they are evaluated in analogues like the Desert Research and Technology Studies (RATS).

If HDC came from the NASA human space flight world, confronting always with real requirements and projects that can evolve in products on board the International Space Station or other Space vehicles, another trend flows in the opposite direction, that regards the intellectual reflection of alternative Design concepts. It is represented by the so-called *Space Architecture* group, which involves both architects and designers intending to challenge themselves on uncommon grounds.

[15]https://www.nasa.gov/feature/habitability-design.

Following the *1st International Space Architecture Symposium*—held by a group of architects and designers, among which also the author, at the World Space Congress in Huston, 2002, according to the *Millennium Charter*, the *manifesto* which we created in that occasion—the definition of Space Architecture has been acknowledged as *the theory and practice of designing and building inhabited environments in outer space* with the motivation of responding to the deep human drive to explore and occupy new places. "Architecture organizes and integrates the creation and enrichment of built environments. *Designing for Space* requires specialized knowledge of orbital mechanics, propulsion, weightlessness, hard vacuum, psychology of hermetic environments, and other topics. Space Architecture has complementary relationships with diverse fields such as aerospace engineering, terrestrial architecture, transportation design, medicine, human factors, space science, law, and art".

The idea behind is that Space Architecture is nothing more than an extension of terrestrial architecture. The goal is the same: to provide a shelter that protects and supports good quality of life for the inhabitants. And as in any architectural project, the designer needs to research and learn all about the environment in which the product is placed. What really changes are therefore the conditions and characteristics of such environment. But when we embrace outer space, the limit of Space Architecture group is to feed a discussion on concept design proposals that not have opportunities to be built as real habitation modules or products for astronauts. The scientific activity is oriented to write papers and books and to organize symposia. Being part of the creation of the Space Architecture group, I presented two papers at the *1st International Space Architecture Symposium* in Huston titled *Designing for Space* and *Space Architecture Education in Milan*, the latter on the experience of a new Master of Science in Space Design that I created and led at the Politecnico di Milano at the end of my Ph.D., over the years 2001 and 2002 and I maintained the relations with the whole group until now. The positive aspect is that the Space Architecture group has increased a network of Space enthusiasts' researchers and professionals aimed to *Design for Space*.

3.4.2 Designer's Perception at ESA

In recent years we saw, there has been a growing evolution in the Space field, triggered by future planned human space missions, which once again propose the desire for exploration as the main engine, now driven principally by the private sector which looks at *space tourism* as a propellant to increase interest and budget. With a permanent settlement on the Moon and a manned landing on Mars as goals to support pioneer ambitions and efforts, a great boost is given to the research, in order to fill the gaps that still make these perspectives impracticable. Following a trend already widespread in several industrial sectors, space agencies are turning to the adoption of *cross-disciplinary teams* as a tool to stimulate innovation.

The group is, indeed, a natural amplifier and accelerator of the creative process, which serves as a *collective brain*, with much higher potential of the single individual brain (Dominoni 2009). Having analysed in depth the group dynamics in the past, I can affirm that the context of cross-disciplinary innovation teams is particularly interesting as a slightly more open door for designers to step in the Space Industry. Recently, I followed the development of a thesis[16] conducted by a student of mine who has participated at the Course of Space Design, *Space4inspiraction (S4I)*[17]—inside the Master of Science in Integrated Product Design, School of Design, Politecnico di Milano—that had the skill, and the good fortune of being chosen as the first and only designer for an internship at the *Spaceship* EAC cross-disciplinary team of the European Astronaut Centre (EAC) in Cologne, a project initially directed by the Italian astronaut Samantha Cristoforetti, after her mission *Future* conducted on board the International Space Station.

The thesis, *A Year in Space: Reflections about the Integration of a Product Designer in Collaborative Space Projects*, was based on the know-how acquired, thanks to my research in 20 years of publications and design for the Space to verify the changes taking place in the light of a methodological aspect that has been taking shape for several years and that concerns the increase of the importance of cross-disciplinary to generate innovation. In particular, this thesis focuses on an interesting setting for the introduction of the diverse perspective of Design in commonly strictly engineering workplaces.

3.4.2.1 The Contamination of Design and Space Exploration

Considering the strong partnership, we consolidated with the European Space Agency (ESA) during these years, overall the last four, developing projects, researches and new educational programmes of Design and Fashion, in which we involved also companies for both Space and private sectors, crossing *multidisciplinary teams*, we wondered if it was possible to *measure* what was the level of contamination today between the world of Industrial Design and that of space exploration, because of the increasing importance of the human being in long-term missions, and also the development of private industry aimed at elite space tourism.

---

[16]*A Year in Space: Reflections about the Integration of a Product Designer in Collaborative Space Projects*, is a thesis by Emilia Rosselli Del Turco; Supervisors proff. Annalisa Dominoni and Benedetto Quaquaro, School of Design, Politecnico di Milano; Co- Supervisors: prof. Paolo Fino, Politecnico di Torino, dr. Aidan Cowley, European Astronaut Centre (EAC).

[17]*Space4InspirAction (S4I)*, which will be treated in this book as a case study, is the first and unique "Space Design" Course in the world recognized and supported by the European Space Agency (ESA) with experts and scientists who suggest and then deepen the project themes, chosen every year together with us, in line with the main objectives of the space agencies and industries' strategic interplanetary programs. *S4I* was created in 2017 by me together with the architect/designer and professor Benedetto Quaquaro, at the School of Design, Politecnico di Milano, inside the Master of Science in Integrated Product Design. While I'm writing we are leading the 4th edition *Space4InspirAction* 2020 in distance, according to the rules for the Covid-19 pandemic.

And overall, I was curious to verify the impact of the contribution of a designer in a multidisciplinary team working for Space and understand what the situation is today in ESA and how the role of the designer in Space scientific environment is interpreted. The thesis aims to discuss the potential benefits unleashed by the introduction of the discipline of design within these cross-disciplinary efforts of the space sector. In particular, a product development of the project named *Flexrack* carried out during the internship at *Spaceship* is used as a discussion ground to reflect on the challenges, the methodologies and skills provided by the work of a product designer in this field as project owner and technical support. Finally, a series of semi-structured interviews makes it possible to compare the views of the industry itself and to formulate some considerations for the future. The conclusions focus in particular on the vision of an ideal collaboration for Design in Space, which aims to make most of the different points of view involved, and on the obstacles identified to its implementation.

We tried to build a state-of-the-art *tailored* to designer point of view in which best practices useful for our purpose could converge: to bring out the importance of Design as a discipline inside *cross-disciplinary teams* for the Space industry. This approach has been recognized and appreciated by ESA teams in various topics, especially in comparison with an engineering approach, and allowed a redefinition of important topics for Design methodologies and innovation as a crossroads of different perspectives.

The interviews provide some recommendations for strategic actions in order to overcome the aforementioned obstacles. A large effort was dedicated to extracting the thematic connected to the overall notion and opinion in the respect of the profession of *designer* in order to lay the groundwork to analyse potential critical points that would jeopardize any present and future collaborative work.

In general, there is a *lack of a precise idea* of what are the approach and the outputs of a designer when coming to technological products. Words used are often generic and redundant, but some topics are recognizable. Among them, none had the experience of actual interdisciplinary collaboration with creative professions, besides short university workshops. Even in these cases, contributions were often separated, and the relationship did not have the time to overcome the initial diffidence and miscomprehension. One of the respondent's reports of having had courses from industrial designers in the context of the International Space University (ISU) Master. Another one, instead, remembers in his answers a previous multidisciplinary collaboration within the automotive industry, where engineers were mainly called to provide feasibility constraints to the *vision* of Design.

3.4.2.2 A Misunderstanding of the Design Role

The results of the interviews show that the *human factors* have recognized a domain of designer, but a clear definition has not given; at the same, the designer is considered able to manage qualitative factors, but it is no evidence of what they are and their impact on the project. The *relation between the human being and the object is missed*, as well as the use of the storyboard to analyse the gesture and the interface dynamics, in order to discover topics not related to the structure or the mechanic but connected with a requirement of flexibility and usability. For example, some modifications of the shelter of the *Flexrack* project that are strongly related to the usability improvement of the product are poorly understood.

The *usability* term when quoted is not connected to the figure of the designer but begins to grow the more generic theme of efficiency. However, during the interviews, it turns out that many errors and problems on the International Space Station (ISS) are attributable to usability problems. The results of this study pushed us to list recommendations to fill the gap of Design in Space inviting to make known the Design and its benefits within this industry. The use of comprehensible language, adopting systems and modes of reasoning, and based on complete case studies and that they have quantitative as well as qualitative assessments it is suggested.

3.4.2.3 Designer Contribution in the Space Industry

A collaborative environment is the most promising context for the introduction of designers in the Space Industry. Nevertheless, collaboration is a hard achievement when coming to the combination of natural scientists, especially engineers, and other disciplines. Design particularly has a set of fundamental differences in terms of methodology and epistemology which makes it difficult for engineering backgrounds to have a clear understanding of its outcomes and processes.

In particular, *Design opposes Engineering* with an unstructured process, an un-standardized result, a broad approach, a deep analysis of the problem, a focus on the qualitative factors and a freest relation with reality's constraints. The latter jeopardizes mutual respect of the expertise which further reduces the openness to actual exchange. Exactly for this reason, though, collaboration is foreseen, in order to ground the designer more on constraints by providing the needed technical knowledge, and to assure a more integrated and diverse perspective next to the engineer's one, to reduce the risk of blind spots and increase the operational functionality.

The designer's role would occur especially in the early phases and with a broad mandate, to *influence* a variety of projects before their implementation. The major obstacles for this to happen are identified in a general conservatism within the Space Industry, in the narrow focus on the most striking and substantial requirements, that makes difficult changes on hyper-tested systems, and in the cultural distance of the engineering system between strongly different Design thinking settings, which makes it difficult to have a conscious appreciation of the possibilities offered by a designer in this field. Also, the management of public money on strict priorities, still connected to pure survival and security of a human space mission make it difficult to introduce new ones related to the *well-being* of the crew.

3.5 To a New Space Era by Design

If the Habitability Design Center (HDC) has the advantage of suggesting, for experience, how the professionalism of the designer is fundamental from the early stages of the design process, because it contributes with its strategic skills to improve the quality of the project, has the big limit of having *self-confined* to be the progressive evolution of the *human factors* approach applied within space agencies and industries by design engineering, without passion and creativity, with a human-centred design attitude that, however, does not integrate the deepest needs of the human being and not even beauty. Also, the complex and multidisciplinary approach, typic of a designer, is here reduced at someone who can visualize ideas to show to customers, a sort of draftsman. In the HDC design process, there no sign of *human needs*, *comfort* and *wellness*. All the effort of the NASA designers seems to dedicate to improve *usability*, which is a direct derivation from the traditional *human factors*, and that identify the methodological approach of engineers to solve topics related to Habitability in Space.

The interest to increase the quality of life, considering also emotional and psychological factors—like Raymond Loewy did convincing NASA to introduce a window to look at the Earth underlining the importance of the entertainment in the Skylab laboratory—and overall, the important role of the beauty to enhance the value of such environments in which humans would live and work in balance—like Galina Balashova use to do through wonderful drawings—seem to be not contemplated. The idea of habitat for NASA corresponds to an aseptic environment, in which prevail smooth surfaces of aluminium easy to clean, covered by racks that contain equipment useful for the maintenance of the station, experiments to be conducted, food and medical supplies for the crew.

Life planned in orbit is made of planned activities, nothing is left to improvise, astronauts follow to the letter very detailed procedures, they do physical exercise not for leisure, but as a necessary countermeasure to counter the negative effects of microgravity, they eat terrible food and perform personal hygiene in extreme conditions, with great difficulty and discomfort, because nothing is designed really for Space, but the majority of the equipment and tools they use are terrestrial, re-adapted for the Space environment.

It is incredible to see how in the most pioneering phase of human space exploration the Russian and American space agencies and Industries felt the need to involve architects and designers in the Design of space stations, including disruptive ideas, like the language of semiotic and colour in Space, the balance between human needs, function and beauty, or the importance of the recreational values for the well-being of the crew. And after more than 50 years of design developments, in which different experiences and data of the various human space missions have been collected, they take a step back, losing these first precious cases that demonstrate the fundamental role of Design discipline in creating habitats in Space, for humans, and not for robots!

Departing from the Shuttle programme and the construction of the International Space Station (ISS), I tried to describe the little consideration of the *habitability* issues, due in part to the lack of understanding by engineering the meaning of Design. I hope that with the setting of the cis-lunar orbital station Gateway and Mars as the next goal in the space exploration roadmap, the competences of Design will be integrated into the whole design process.

Despite last decades saw an incremental attention towards space manned missions, and their requirements with the final aim of really enabling a permanent human presence in Space, the involvement of Design approach in Space is far to be completed. The need for long-term travels and, consequently, separation from Earth placed the focus on the dynamics triggered by humans' involvement in a mission's design to start to learn more about how these would behave and interact in such scenarios.[18] Furthermore, the progressive enlargement of the group of people having the possibility to experience Space to scientists, researchers and even tourists, over long-time trained astronauts, opens to the need of reconsidering the habitability issues related to life in Space environments.

In the light of the above, in the Space field, the necessity appeared to go beyond the pure technical approach led by engineering team and slowly open to other competencies. Design, among them, represents a point of interest, thanks to its experience with the integration of users' needs and behaviours with the productive and technical worlds.

3.5.1 An "Italian Design Style" Spaceship

Design becomes indispensable to create new environments and objects designed specifically for Space that can *turn the limit, as is the microgravity, into an opportunity*. And perhaps, it would be advisable to think also of environments for the amusement and the entertainment of the crew, like we have been asked by Thales Alenia Space (TAS) which has commissioned us a recent project for a new spaceship looking at a *recreational space*, with more room for astronauts' entertainment. The assignment concerned *Innovative Concept Design Solutions aimed to Emphasize the Italian Style by Design* supporting the programme for a *New Orbital Infrastructure Recreational/Habitable Configurations* and specifically for *Window Design in Pressurized Modules* (*task 1*) and *Recreational/Habitable Module Interior Design* (*task 2*) introducing also proposals of fashion and products design elements in the interiors habitat under definition of TAS in Italy.[19] We applied

[18]Evan Twyford in the video *Industrial Design for the Outer Space*, Vice Media 2012, NASA Archives.

[19]I led and developed the project of the new spaceship together with Benedetto Quaquaro identifying, according to Thales Alenia Space, *Preliminary Design Solutions for the Support to New Orbital Infrastructure Recreational/Habitable Configuration* inside our *Space4InspirAction Lab* at the Design Department of Politecnico di Milano.

design methodologies as *User-Centred Design (UCD)* and *Use and Gesture Design (UGD)*, together with ergonomics aspects, to increase the comfort and the efficiency of the crew in confined environment and microgravity condition optimizing crew space usability and subsystem arrangement through flexibility and re-configuration. Our proposals concerning *Concept Design Solutions* for both external and internal configurations and volumes of a new Space Station were focussed on the two tasks given: *Window Design and Recreational/Habitable Module*.

Concerning the *task 1*, our purpose was to design various solutions injecting style elements to the existing technical solution conceived by TAS to *build a strong identity* of the new Spaceship focussing on *recognisability* (also from outside the Spaceship) modifying panes alternative shapes and colours according to the materials' requirements. The goal we wanted to achieve was to transform the Spaceship in a precious object, a *Jewel in Space*, in the dark sky, designing a new configuration alternating round and oval windows as precious stones creating a "ring" to highlight the window light stands out against the dark sky, if black, or the aluminium modules softly, if gold, or strongly, if red.

Concerning the *task 2*, the study of the internal accommodation of the recreational/habitable areas was focussed on four topics: re-configurability, free room optimization, illumination adaptability, stowable functional furniture (e.g. working/lunch desk, restraints); the innovative design solutions related to fashion and products design elements in the interior habitat explored various possibilities among which relaxing couch, privacy feeling enhancement (e.g. isolation, including acoustic noise), handrails, foot restraints and stowage solutions focusing on aspect of technical feasibility and in compliance with the constraints/requirements given by the preliminary concept provided by TAS. Imagining the *Jewel in Space* from inside, it reveals an immersive experience into the dark Space, the feeling we want to transmit was *entering the window*, surrounded by glass—sliding into the inter-space created by the round window—and "looking at the stars lying on the sofa".

We have re-designed completely both the exterior skin, including the windows, that alternate different shapes and create the idea of a transparent and continue surface, and the interior of the habitation module: an acoustic textile padding able to reduce noise and increase acoustic isolation covers the structure of the windows, as well as the interior of the module, creating *sofa rings* and giving a smooth feeling like a hug.

The project of a *New Orbital Infrastructure Recreational/Habitable Configurations* was related to the request of Thales Alenia Space to find "Innovative Concept Design Solutions aimed to Emphasize the *Italian Style* by Design". The inspiration which has driven this not easy task was made by three keywords: *Italian Design—Sixties—Space Age*.

"Re-configurability" was the key to gain more internal space and flexibility. The racks on the International Space Station (ISS) are replaced by easily removable cylindrical volumes that can slide and transform themselves into "tailored chaises longue" for a *free room optimization* (Fig. 3.15). The cylinders run on tracks serving as soft surfaces of support, as well as shelves, or removable containers. The internal space of cylinders it is foreseen for stowage of equipment and tools.

Fig. 3.15 *Preliminary Design Solutions for the Support to New Orbital Infrastructure Recreational/Habitable Configuration* designed by Annalisa Dominoni and Benedetto Quaquaro for Thales Alenia Space: the racks of ISS are replaced by easily removable cylindrical volumes, while light changes intensity and colour following circadian rhythms with the aim to change the perception of the environment according to the various activities and create a space of relaxation and wellness. Credits by the authors

Flexible restraints to maintain *softly* the body posture on the cylinders/chaises longue are foreseen, as well as accessories for working, having lunch, resting and so on. Female Velcro® shapes are distributed along the whole soft surface to allow crew moving or maintaining the body posture thanks to male Velcro® socks.

The re-configuration of the internal spaces has been given to gain more free volume and flexibility according to the different activities and needs of the astronauts: working, relaxing, gathering, reading, playing and listening to music, resting, looking outside the windows, organizing meeting and lunchtime.

Light changes intensity and colour following circadian rhythms with the aim to change the perception of the environment according to the various activities, while the overall feeling is of relaxation and wellness (Fig. 3.15).

We put *wellness* as focal point of the whole project creating a *sensorial environment* in which all senses were stimulated, to contrast the lack of natural stimuli in Space. In order to reach the illumination *adaptability*, we imagine *LED Rings* for diffused light that change intensity and colour in relation to the various activities on board and integrating circadian rhythm system with chromotherapy: for examples, white-gold light during daily life activity, white-blue light when a relaxing

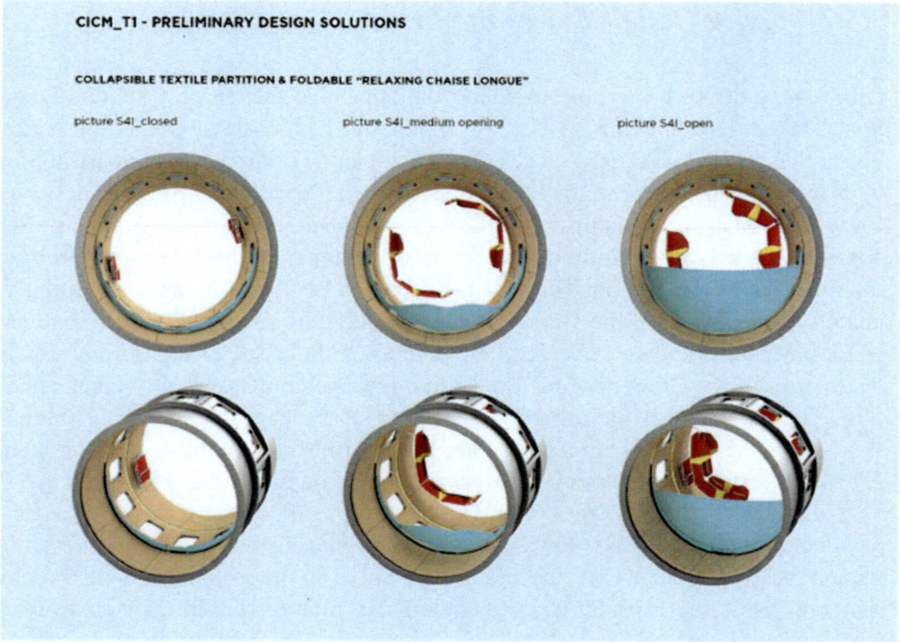

Fig. 3.16 Among which specific crew system/outfitting items and tools to be integrated into the interior of the *New Orbital Infrastructure Recreational/Habitable Configuration*, designed by Annalisa Dominoni and Benedetto Quaquaro for Thales Alenia Space, there is a collapsible textile partition inspired by the *gennaker snuffler* used in sail competitions, with the aim to enhance space re-configuration for crew privacy and socialization purpose, and a foldable relaxing chaise longue to provide a support for rest of the activities. Credits by the authors

environment is required, gold light for gathering, recreational activities and eating together or orange light for fitness activity that implies to be very active and energetic. Removable personal lights to be fixed where needed inside the space station were also foreseen. In addition, we propose to introduce *light walls* to divide virtually the interior space of the module introducing the use of Artificial Reality (AR) for fitness activities and immersive experience into nature.

The whole project has foreseen the design of the environment and the proposal of specific crew system/outfitting items and tools to be integrated into the interior with the aim to enhance space re-configuration for crew privacy and socialization purpose, as a *collapsible textile partition* (Fig. 3.16) made of innovative acoustic textile inspired by the "gennaker snuffler" used in sail competitions, and a *foldable relaxing chaise longue* to provide a support for relax and rest activities.

If we compare the various projects and the "space" given to Design to express its competences and skills in this chapter, we can easily see the evolution of Space Design contribution in the Space Industry.

3.5.2 Space Design Drives the Private Space Industry

On 30 May 2020, I was able to attend the launch of the SpaceX's Crew Dragon Spaceship and the docking on 31 May, after about 19 h of travel to the International Space Station (ISS). A great success for America that returns after 9 years in Space with a new launcher, Dragon 9, which replaced the Shuttle. SpaceX developed Crew Dragon in partnership with NASA's Commercial Crew Program. In 2015, NASA selected SpaceX to pursue crewed commercial flight to and from the International Space Station. This was the agency's first mission order for SpaceX to launch astronauts from the United States. SpaceX, the private aerospace company of Elon Musk, is known to be very attentive to the futuristic Design of its vehicles: the interior of the Crew Dragon[20] looks like it came straight out of a science fiction film. The black-and-white ultramodern design of the capsule is visually stunning. Aside from its seven seats, which are made from carbon fibre and the Italian Alcantara textile, the interior of the capsule has a number of key features.

SpaceX wanted a voyage in Crew Dragon to be an *enjoyable experience* for its passengers. It has four windows and a sleek digital display that will show information ranging from where the craft is in Space to information about the environment inside the capsule. Crew members can monitor the interior environment with an environmental control and life support system, setting the on board temperature between 18 and 27 °C. The private Space enterprise has also designed spacecraft to be fully autonomous, though astronauts on board will be able to monitor and control the craft using facilities on board the Crew Dragon. The same focus on Design has been placed on spacesuits for astronauts, designed by Hollywood costume designer Jose Fernandez, who combine aesthetics and technology. His specialization in *superhero wardrobe* has certainly influenced the development of suits for SpaceX, in which the alternation between white and dark grey, ovoid helmets, high boots and geometric lines that define the details give life to a comfortable and futuristic silhouette.

It is therefore clear that Design takes on strategic importance when dealing directly with the private space industry, especially if guided by visionaries who are not content to ensure security and technological reliability of the vehicles, but they want to transform the *experience* of the human space flight's into something magical and unforgettable, they want to give shape to a dream.

References

Belew LF (1977) Skylab, our first space station. George C. Marshall Space Flight Center, Scientific and Technical Information Office. National Aeronautics and Space Administration (NASA). Washington, D.C. https://history.nasa.gov/SP-400/contents.htm

[20]Crew Dragon Interior https://www.youtube.com/watch?v=xjSb_b4TtxI&feature=youtu.be.

Clark P (1988) The Soviet manned space program. An illustrated history of the men, the missions, and the spacecraft. Orion Books, London

Compton WD, Benson CD (1983) Living and working in space. A history of Skylab. The NASA History Series, National Aeronautics and Space Administration, Washington, DC. Scientific and Technical Information Branch, NASA. https://ia800704.us.archive.org/20/items/nasa_techdoc_19840017669/19840017669.pdf

Cohen M (1995) The suitport's progress. In: Life sciences and space medicine conference, 03–05 Apr 1995, Houston, TX. American Institute of Aeronautics and Astronautics (AIAA)

Connors MM, Harrison AA, Akins FR (1985) Living aloft. Human requirements for extended spaceflight. Scientific and Technical Information Branch, National Aeronautics and Space Administration (NASA). Washington, D.C.

Dominoni A (1998) IVA clothing support system. In: Proceedings of space exploration and resources exploitation, ESA ExploSpace Workshop, 20–22 Oct 1998, Cagliari, Sardinia. European Space Agency (ESA), Noordwijk, WPP-151, pp 0015.1–0015.4

Dominoni A (2002) Industrial design for space. Silvana Editoriale, Milan

Dominoni A (2005) VEST. Clothing support system on-orbit validation. In: SAE technical paper series, ISSN 0184-7191, Proceedings of ices 35th international conference on environmental systems, Rome, 11–14 July 2005. SAE Society of Automotive Engineers, Warrendale, Pennsylvania, USA

Dominoni A (2009) Creative network systems for innovation. In: Proceedings of INTED 2009, international technology, education and development conference, Valencia, 09–11 Mar 2009. Association of Technology, Education and Development (IATED), pp 002572–002580

Dominoni A (2015) For designers with their head beyond the clouds. Politecnica. Architettura, Ingegneria, Scienze, Maggioli Editore, Milan

Evans B (2012) At home in space: the late seventies into the eighties. Springer.

Gazenko OG, Shumakov VI, Kakurin LI, Katcov VE, Chestukhin VV, Nikolayenko EM, Gvozdev SV, Rumyantsev VV, Vasilyev VK (1982) Effects of various countermeasures against the adverse effects of weightlessness on central circulation in the healthy man. Aviat Space Environ Med 53(6):523–530

Johnson CC (1974) Skylab experiment M487, habitability/crew quarters. In: Paper 133 of the Skylab results. Proceedings, vol 1, 20th American Astronautical Society Meeting, Los Angeles, 20–22 Aug 1974

Jordan PW (2000) Designing pleasurable products. An introduction to the new human factors. Taylor & Francis, London

Loewy R (1981) Living in space: NASA projects Apollo, Saturn, Skylab, Shuttle. Sotheby's, New York

Mead GH, Peercy RI, Raasch RF (1985) Space station crew safety. Alternatives study—final report, vol V—space station safety plan. NASA Contractor Report 3858. Contract NAS1-17242, June 1985

Meuser P (2015) Galina Balashova. Architect of the Soviet space programme. Dom Publisher, Berlin

Moore D, Bie P, Oser H (1996) Biological and medical research in space. An overview of life sciences research in microgravity. Springer, Berlin, Heidelberg

Musso G, Gaia E, Battocchio L, Colford N (1996) Space hardware for long duration flights: Tool Box and Elite-S. In: 47th international astronautical congress (IAC), Beijing, China

Payne FA (1960) Work and living space requirements for manned space stations. In: Proceedingsof the manned space stations symposium, Los Angeles, 20–22 Apr 1960, pp 100–103

Peercy RI, Raasch RF, Rockoff LA (1985a) Space station crew safety. Alternatives study, vol I—final summary report. NASA Contractor Report 3854. Contract NAS1-17242, June 1985

Peercy RI, Raasch RF, Rockoff LA (1985b) Space station crew safety. Alternatives study—final report, vol IV—Appendices. NASA Contractor Report 3857. Contract NAS1-17242, June 1985

Raasch RF, Peercy RI, Rockoff LA (1985) Space station crew safety. Alternatives study, vol II—threat development. NASA Contractor Report 3855. Contract NAS1-17242, June 1985

Rockoff LA, Raasch RF, Peercy RI (1985) Space station crew safety. Alternatives study—final report, vol III—safety impact of human factors. NASA Contractor Report 3856. Contract NAS1-17242, June 1985

Souza K, Etheridge G, Callahan PX (2000) Life into space. Space life sciences experiments. Ames Research Center, Kennedy Space Center 1991–1998

Stuster J (1986) Space station habitability. Recommendations based on a systematic comparative analysis of analogous conditions. NASA Contractor Report 3943. Scientific and Technical Information Branch

Taylor TC, Spencer JS, Rocha CJ, Khan E, Cliffton E, Carr C (1987) Space station architectural elements model study. NASA Contractor Report 4027. Purchase Order A211774C, Jan 1987

Taylor GJ (2008) Designing for space: Core77 visits NASA's industrial design team. https://www.core77.com/posts/11565/Designing-for-Space-Core77-visits-NASAs-Industrial-Design-Team-by-Glen-Jackson-Taylor

Wichman HA, Donaldson SI (1996) Remote ergonomic research in space: spacelab findings and a proposal. Aviat Space Environ Med 67(2) (Alexandria, VA)

Chapter 4
Space Design Between Research, Project and Education

4.1 Five Case Studies of Space Design

In this chapter I have chosen five case studies from my work of designing,researching and teaching Space Design to support the previous three chapters with the aim to demonstrate the advantages to integrating the Design discipline in the Space industry. The case studies are very different, but follow a chronological order; according to the development of my research path, and overall, they are all *first pilot projects* of Space Design, and are pioneers of a discipline and a methodology that I have tried to affirm throughout my career, now more than twenty years.

My intention is that the results of these projects could highlight the role of Design for Space together with a well-deserved recognition in the space scientific community of this new discipline.

My intention is that the results of these projects could highlight the role of Design for Space together with a well-deserved recognition in the space scientific community of this new discipline.

In addition, all the case studies demonstrate the importance of a multidisciplinary approach in which Design emerges as a *bridge* between science and beauty, able to speak the various disciplines' languages as well as manage working groups, made of different backgrounds and know-how according to the nature of the projects, which could not only embrace the area of a clothing support system to perform activities in microgravity, but also equipment and tools to facilitate the various activities on board, lighting systems, lunar or orbital bases and so on.

The designer should interpret—by means of the formal and symbolic qualities of its projects—people's needs, expectations and desires in relation to the environment, and at the same time elevating the degree of their general well-being, the functionality of the activities and the usability of interface's facilities and instruments for astronauts, to achieve better goals that directly affect the success of human space missions. Not last, these projects show how fundamental is the involvement of the companies, both in the space industry and in the private sector,

A. Dominoni, *Design of Supporting Systems for Life in Outer Space*, Research for Development, https://doi.org/10.1007/978-3-030-60942-9_4

to design objects, equipment and habitable systems that should be at the same time innovative and feasible.

4.1.1 VEST. Project and Experiment on Board the ISS

The first case study describes, in fact, two case studies, but joined by the same topic, an innovative integrated system of garments designed for the crew activities, *VEST* and *GOAL*, that were born by a project research financed by the Italian Space Agency (ASI) in 1998, and that I developed as a feasibility study with the aim to design models and prototypes to be tested for two experiments on board the International Space Station (ISS) in which I was the Principal Investigator: *VEST, Clothing Support System for the International Space Station*, was performed during the Italian *Marco Polo Mission*, in April 2002, and *GOAL, Garments for Orbital Activities in weightLessness*, during the *Eneide Mission*, in April 2005.

The two experiments *VEST* and *GOAL* are the tangible results of my intuition of an innovative integrated system of garments to be used during the human space flight missions with *dedicated* Space items, for microgravity conditions and confined environment, when performing Intra-Vehicular Activities (IVA), that increase comfort, efficiency and wearability.

Aim of the experiments was to demonstrate that the provision of a new integrated system of garments to the crew would not only increase their well-being through the use of specific fabrics for confined environments, able to improve thermoregulation and body hygiene, but also tailored models designed for the Neutral Body Posture (NBP), assumed by the human being in microgravity, that integrate particular repositionable pockets and restraints to move better through the interiors of ISS as well as to fix the body when settlement activities are required.

Also, not expecting to use additional time for the experiments, the clothing were worn during normal daily activities, work and rest, allowing to collect useful data with less impact on the crew schedule, and to prove that in general with *VEST* and *GOAL* equipment less mass and less volume were assured.

This case study also describes an example of spin-in, in which I applied cuts and seams taken by the Earth snowboard clothing (Dominoni 2002) that then I translated to the new integrated system of garments because I found the posture performing this sport is very similar to the compact NBP assumed by the body in microgravity with knees towards the shoulders.

I am very proud to affirm that *VEST* and *GOAL* are the only two example until now in which a space agency, in this case ASI, sponsored and supported the realization of a *fashion design project*, and overall, accepted to collaborate with a private company, Benetton Group, that I had involved in the project because they specialized in sport and technical garments. It was my first *gesture* to demonstrate that also the private sector, if well driven, could be involved in the space programmes. We could consider that gesture as one of the first strategic action driven by Design that now we would call it *new space economy*.

4.1.2 Couture in Orbit. A Capsule Collection for ESA

The second case study presented is the project *Couture in Orbit*, an experience of research and teaching that I led with a group of selected students at the School of Design, Politecnico di Milano, on the invitation of the European Space Agency (ESA) in 2016, with the aim to create a *capsule collection* inspired by Space scenario, *using ESA technologies*, and designed to be worn every day on Earth.

Compared to the two previous case studies, in which *VEST* and *GOAL* are both projects of a new integrated system of garments designed to be worn by astronauts on board the ISS, and in which Design serves space environment (*Design for Space*) generating a spin-in, *Couture in Orbit* is the reverse case study, in which Design is inspired by Space, and is pushed to imagine new applications for space technologies to integrate into terrestrial garments (*Space for Design*) generating spin-offs.

Working together with the Italian partner of ESA for the Technology Transfer Solutions, Rina Consulting, we have carried out extensive research into ESA space technologies patents to try to interpret and transform them looking at the social needs, with the aim to induce new behaviours driven by technology.

Our principal aim was to find relations between life in Space and life on Earth connecting the two environments and considering not only technological spin-offs or spin-ins, but also emotional language and usability.

We transformed inspirations into shapes and textures by looking at astronaut postures and movements in microgravity and considering their activities in Space in relation to the objects they use and the environment they live, in order to imagine new gestures, interpret new needs and generate new ideas.

Couture in Orbit today remains the first and only case study in which a space agency, this time ESA, decided to use the design and fashion language to communicate to a huge public its role with the aim to celebrate space exploration and research exploitation and their impact in many fields of our life.

4.1.3 Space4InspirAction. A Space Design Course with ESA

The third case study is *Space4InspirAction (S4I)* that I created in 2017 together with Benedetto Quaquaro and now is at its 4th edition 2020. *S4I* is the 1st and unique international Course of *Space Design* in the world, recognized and supported by the European Space Agency (ESA) through experts and scientists who suggest and then deepen with us the project themes in line with the objectives of the space agencies' strategic programmes. The Course *S4I* is the principal design laboratory in the second semester, inside the Master of Science in Integrated Product Design, at the School of Design, Politecnico di Milano.

The collaboration between the *S4I* Space Design Course and ESA is not a spot project, but the beginning of an ongoing agreement about an *academy—research—*

design process to ensure the development of innovative ideas in which the contribution of design discipline is recognized by ESA as fundamental and strategic to generate *disruptive visions* of habitability in extreme environments and equipment solutions to increase the *wellness* in Space.

Thanks to ESA, we are able to compare our design vision with real conditions and requirements and find innovative solutions designed for Space, but that can be also transferred to daily life on Earth becoming spin-offs.

Space allows students to develop creativity with intense visioning activities by confronting their projects with confined and microgravity environment, that is not part of our common experience. The unusual and extraordinary environment offers the students the possibility to think *out-of-the-box*, like astronauts experience looking at the Earth from another point of view, outside the atmosphere, together with the chance to increase the creativity and the ability to imagine and design new objects and tools starting from a perspective completely unknown.

Through a new methodology that I have developed and called *Use and Gesture Design* (UGD), the Design role is not limited to creating objects, environments, architectures, but also *new behaviours and gestures* of astronauts in relation to the use of designed objects, environments and architectures.

4.1.4 Fashion in Orbit Versus Space Fashion Design

The fourth case study is the unique spin-off generated by *Couture in Orbit*. Looking at the extraordinary results of the project presented at the London Science Museum on May 2016—during an event including a catwalk and an exhibition of the materials and technologies used for the capsule collection—which have generated over 20 million media feedback responses, and we decided (Benedetto Quaquaro and I) to continue the experience together with ESA, creating a specific Higher Education Course, called *Fashion in Orbit*, at POLI.design (the Consortium of Politecnico di Milano) in February 2017, based on the same idea of *Couture in Orbit*: a capsule collection of garments to be worn on Earth that integrates Space technologies and innovative materials.

Fashion in Orbit was not a curricular Course scheduled inside a Master of Science like *Space4InspirAction*, but was open to young professionals who wanted to develop innovative products in fashion tech and smart textiles, inspired by space and technology research, in order to find new applications and spin-offs for materials and textiles driven by attractive solutions of garments, which are able to increase the value of the technology itself. *Fashion in Orbit* pillars included fashion, technology and space inspiration.

We involved specific companies of technology focussed on sensor systems and data collection, together with many other companies of the private sector coming from smart textiles and luxury fashion fields. Colmar, an Italian brand of sportswear, very known for the ski clothing, was our main sponsor, while the other companies—Alcantara, Caimi Brevetti, Coats, Limonta, Luxury Jersey,

Omniapiega, Sitip, Soliani, Thermore—supported our projects supplying materials, fabrics, accessories and fabric processing systems to realize the capsule collection.

Our strategy was to put together companies coming from distant fields of application and know-how, as the new acoustic textiles product made by Caimi Brevetti—which integrates a strong scientific research on the sound propagation—with Omniapiega—which is dedicated to producing pleats of haute couture fabrics in silk—with the purpose to create innovation. This strategy of putting around a table distant knowledge—observing products and productive processes which coming from areas that we never imagined integrating and trying to generate new connections and thoughts asking ourselves *what if?*—we found it is a very good methodology to produce disruptive ideas and innovative products.

All the conceptual works emerged at the end composing the capsule collection of *Fashion in Orbit* represented different identities and backgrounds, as the companies involved, and shared an intimate relationship with the cosmos, which was expressed through garments with new languages full of emotion, feeling and poetry, mixing fabrics, materials and technologies in unusual and surprising ways.

A year later, in 2018, we extend the area of application of fashion to product design, to give more possibility to innovate crossing different fields, and we called the 2nd edition *Space Fashion Design* transforming the formula of the Course for professionals to a contest in which we worked focussed on a unique brief defined together with the private company and ESA.

4.1.5 *Moony. A Lunar Base in Lava-Tubes for Igluna ESA_Lab@*

The fifth and last case study I present in this book confirms the overture and the interest of the European Space Agency (ESA) to improve cooperation across Europe by building an international and multidisciplinary platform through the collaboration between different disciplinary perspectives developed by different universities and research centres. The principal aim of this new project is to create an educational environment able to inspire a potential *future generation of Space experts* and to analyze the future development of technologies being tested to discover possible new potentialities for human space exploration.

I was invited in September 2018, together with Benedetto Quaquaro, to be part of *Igluna*, the pilot project of Esa_Lab@ coordinated by the Swiss Space Center and supported by the European Space Agency (ESA): a multidisciplinary platform to create a *space habitat* with the aim to demonstrate how to sustain life in an extreme environment using space technologies for disruptive innovation and cross-fertilization. Teams of students from all over Europe, supervised and supported by their academic institutions, were called upon to develop a series of modular demonstrators inspired by the context of a lunar habitat.

The Politecnico di Milano was represented inside *Igluna* project only with our Design Department. Having the role of *designers* we were asked to *give shape* to the whole lunar base—that was to be built in the *lava tubes*, under the Moon surface, to protect the human settlement principally by cosmic radiations and meteorites—integrating into the internal configuration of all the other projects (a part of robotic arms to help the realization of the lunar base in situ, greenhouse systems for extreme environments as well as *boxes* containing experiments for life science) that should have converged in a final exhibition of the results.

Particularly, our *Moony* project, being a link between the actually built habitat and its vision on the Moon, tried to picture a habitat and mission development that could appeal imagination, but restraining creativity to plausible technologies and operations, based on solid case studies. There are several issues addressed, such as how to build and place a housing module safely in an extreme context—responding to requirements that can encourage the adaptation and comfortable stay of the human beings—or which king of technology to use to build the base—integrating in situ materials, as regolith, with collapsible inflatable structure—that requires research, experimentation, tests and in which demonstrators result to be very useful.

The *analogue*, as the demonstrators of space environments reproduced on Earth in territories with similar characteristics are called, was completed inside the swiss glacier of Zermatt, the Matterhorn Glacier Paradise, and has been inaugurated at the end of June 2019 to test all the prototypes that belonged to the various experiments. The 1st edition of *Igluna* 2019, brought together 19 teams from 13 universities in 9 different European countries.

The multidisciplinary nature of *Igluna* project covered a full range of topics, going from the habitat conception and construction to life support systems, communication, navigation, power management, as well as science and human well-being. If we imagine to build a livable environment—especially one far from the Earth—we should take into account all aspects of human life, that includes food production, energy and oxygen, the construction of shelters, the provision of instruments and the enabling of communications. Teams from different universities were either mono-disciplinary or cross-disciplinary, but they worked in the context of a multidisciplinary project where different contributions required to be integrated developing oxygen and electricity production processes, a mechanism to cultivate algae, a system to use urine to fertilize plants, sports facilities, a laboratory, a system of communication and control, a robot for digging paths in the ice and more. Expertise involved ranged from a different kind of engineering, from aerospace, biologic, electronic, robotics to computer science, geology, architecture and design disciplines.

4.2 VEST. Projects and Experiments on Board the ISS

In April 1999, the area for scientific research of the Italian Space Agency (ASI) organized the national workshop on the *Scientific Use of the International Space Station (ISS)*, which was attended by about 200 participants from the Italian scientific and industrial community. On that occasion, the opportunity was created for a better knowledge of the resources and support that the ASI could provide and it was clear that there was considerable interest from the scientific community in terms of programmes and proposals, although not evenly distributed across the four thematic areas. One year after this initiative, also in the light of the restructuring of the ASI, the workshop on *Scientific and Technological Use of the ISS* was organized at the astronomical observatory of Capodimonte in Naples on 17, 18 and 19 April 2000. The aim of the workshop was to stimulate and address scientific and technological proposals and projects, national and international collaboration, and verify the needs of the community in the different stages of proposal, design and implementation of programmes, also in view of the initiatives of the ASI, for the development of operational and engineering centres in support of the scientific community. The participation of those interested in the technological use of the International Space Station (ISS) was one of the innovative elements of this workshop and represented a moment of better integration between scientific and technological programmes in terms of future collaborations.

The achievement of these objectives has been through presentations in plenary session followed by debates and meetings of working groups on the four thematic areas—science of engineering, science of Earth, science of the universe, science of life—during which it was possible to discuss the status and requests of all ongoing projects selected within the European Space Agency (ESA) or international for human space missions on ISS, Shuttles, rockets, parabolic flights, etc., so as to allow room for a final discussion.

I participated in the science of life's thematic area with my *VEST* project, an innovative integrated system of garments, selected for an experiment on board for the ISS, in which I proposed the development of a new clothing system designed for life and work in Space that which foresaw tailored made garments, included the search for new cuts, seams and tissues suitable for confined environment and microgravity, and the integration of prostheses in the garments to facilitate movements and Intra-Vehicular Activities (IVA) of the crew and make them more comfortable for the benefit of greater overall well-being.

Considering that the state of the art of the clothing on ISS does not provide specific garments designed to be worn in microgravity and confined environment, and *VEST* project represents a unique case study in which astronauts have been able to experience to feel more comfortable, efficient and clean—thanks to the use of antibacterial and thermoregulating fibres as well as the attention I have given to new parameters of functionality and wearability according to the Neutral Body Posture (NBP) assumed by the body in Space.

4.2.1 Proposal for the Development of an Integrated Clothing System VEST

The *VEST* project gives me the opportunity to report in this book a testimony, quite rare in the field of Design, of how the various phases of development of the project took place: from the first proposal, the *feasibility study*, up to the realization of the *prototypes*[1] of the garments collection, and the subsequent *experiment* conducted in orbit on board the International Space Station (ISS). During the design and development process, the importance emerges, which I stress several times also in the other case studies, of creating multidisciplinary groups made up of experts and companies, both in the space and private sector, to complete all the parts that make up such a complex project.

For a designer, it becomes essential to be able to *build a network of interlocutors*: starting from space agencies, the Italian Space Agency (ASI) and the European Space Agency (ESA) in this case, which are a fundamental reference for the sponsorship and the support of the project; continuing with an industry specialized in habitable pressurized modules such as Thales Alenia Space, which is fundamental for the *spatialization* of the projects, and which has dealt precisely to spatialize my prototypes; and finally integrating companies that possess a great know-how, specific for terrestrial products, such as in this case Benetton Group – which I chose for its international recognition as a fashion manufacturer, in particular sportswear, and for the research and development center within it, specializing in advanced textiles and nanotechnology, which could support me in the development of the most suitable fabrics and garments for the confined environment and microgravity—as well as Sorin Biomedica Cardio, for the development of monitoring systems made of micro-sensors and external tools to manage signals and collect data.

For the Italian Space Agency (ASI), it was the first time in which a proposal integrated industries of the private sector, instead of Space industry and they appreciated the innovation in the process, and the strategies related to the design thinking approach that I applied, from the beginning of the first ideas to the realization of the experiment on ISS with the astronaut Roberto Vittori, who was involved both in the *VEST* and *GOAL* projects, the second one the implementation of the first project, following the results I achieved.

4.2.1.1 Aims of the Project

The principal purpose of the project *VEST* was the development of a new integrated clothing system designed for life and work in Space, which involved the search for new tissues, and the integration of prosthetics tools in the garments to facilitate movements

[1]In the Space industry, usually the *prototype* coincides with the *first product of the series* considering that we produce only unique pieces realized by engineers' teams which we can consider *high tech artisans*.

and Intra-Vehicular activities (IVA) of the crew. The clothing system, already presented in the paper *IVA Clothing Support System, Explospace, Workshop on Space Exploration and Resources Exploitation, ESA* (Dominoni 1998) provided the development of a collection of clothes for the different activities to be carried out on board the ISS: clothes for work, for rest and for physical activity, in addition to underwear.

The main objectives were to improve the overall well-being of the crew, but also, which at the end it is the same thing, to reduce the logistical demand caused by the number of astronauts and the duration of the human space missions, which requires or have many changes or to be able to wash dirty clothing. The clothes could have then been used to help reduce some unpleasant effects of the presence on board, such as absorbing unpleasant odours and helping the skin to release its dross, both as liquids (sweat) and as solid products (skin flakes). This would have also led to an improvement from the point of view of reducing microbiological contamination on board. The project pushed to increase.

1. *the level of comfort and efficiency* currently achieved, facilitating the movements and the operational activities of the astronauts through a range of equipment and prostheses, incorporated in clothing, which allow the anchoring of the crew to the internal structure of the modules of the International Space Station (ISS) and the transport of tools and objects that are fixed on clothes;
2. *the aesthetics and wearability* of the garments, currently, astronauts wear clothes that they bring to Earth and that are not suitable for the particular environmental conditions on board. I was convinced that the concept should be changed and that it was important to design *style, cut and fabrics suitable for microgravity*, considering the particular movements and postures, not static, which are created inside the pressurized modules. In addition, a study of smart and performing fabrics was needed—to meet the needs of the confined environment with technical garments—an in-depth analysis of shapes and colours—in order to diversify the items and provide a wider choice for astronauts, while taking into account environmental requirements and physiological changes related to microgravity—for the benefit of a higher level of well-being and a better quality of life and social relations on ISS.
3. *body hygiene and thermal stability*, through the use of fibres and tissues we are able to perform the function of thermoregulators, absorbing excess heat and releasing it when the body cools, but also able to limit the harmful effects of perspiration by transferring excess sweating to the outside of the garment;
4. *the integration of prosthetic elements in garments* able to amplify movements in microgravity and facilitate anchoring, ensuring the stability of posture during the various activities on board, and solutions to fix *small tools* directly on the body;
5. *control of the vital, biological and physiological functions* of the crew, through the monitoring of body values such as temperature, heart rate, circulation, breathing, physical stress, fatigue, etc. without requiring the active participation of astronauts. In particular, this project area provided for the presence of sensors inserted inside the garments, but also a study on the possibility of miniaturizing the sensors to insert them into the same fibers of the fabric.

4.2.1.2 Reasons for Using the ISS to Test VEST Project

The project of the new *Integrated Clothing System VEST* provided the indispensable condition to be tested on the International Space Station (ISS) to verify the objectives described above (Dominoni 2001a, b):

1. *equipment and prostheses incorporated in clothing*—which allow the anchoring of the crew to the internal structure of the modules of the International Space Station (ISS) and the transport of tools and objects that are fixed on clothes—required precise and detailed experiments and verifications that cannot be replaced nor with the tests with Neutral Buoyancy Facility (NBF) in the pool—the water being more viscous than the air does not allow a correct assessment of the stability obtainable: in water, it will be more and more stable—neither with parabolic flights, because the microgravity conditions are excessively short;
2. *the style, the cut and the fabrics suitable for microgravity cannot return the same effect in the pool*—in the presence of water it creates adhesion and the fabrics become heavier—and the parabolic flights' results, as in the previous case, are too short to verify the various postures and movements that are determined by the zero-g conditions;
3. *the thermoregulation and drainage characteristics of clothing fabrics* must be tested on board the ISS to verify, for example, the ability of the garment to carry liquids on the outer surface, without causing them to disperse into the environment in the form of water, but also that the transport phenomenon is repeatable in the absence of weight, obviously impossible to verify, for example, in the pool;
4. *the sensors* should be tested on board the ISS, during physical and work activities, because in the pool would create problems of data transmission from the skin to the sensor itself—little adherence of the sensors to the skin due to the presence of water—and with parabolic flights, there would be no time to monitor the various functions described above. This is a problem that is not easy to solve even in orbit, given the difficulties with a similar concept during the mission EUROMIR '95 for experiments in human physiology and respiration.

4.2.1.3 Technical and Technological Aspects

The project consists of several parts, each of which requires specific research and technological insights to meet the objectives in relation to the particular conditions of microgravity as well as the internal configuration of the pressurized habitable modules of the International Space Station (ISS).

For fabrics (I involved Benetton Group for the expertise in textiles and sportswear collections), natural and synthetic fibre technologies would have been used with applications of Thermolife (thermoregulator), Purelife (antibacterial), Sunlife (ceramic material), Drylife (rapid evaporation of water), anti-slip, anti-rip inserts, Easy Care (anti-wear treatment), in different percentages and shapes depending on the garment. The characteristics of softness, lightness, elasticity and breathability of clothing were obtained, thanks to the type of construction of the mesh and the resulting weight.

The absence of seams, obtained through a new technological process, further improves the fit of the garment *eliminating the problem of differentiated sizes*. The development of sensors to monitor the various biological and physiological functions of the crew (Sorin Biomedica Cardio participated to the project researching the most suitable solutions) provides technologies related to stimulation catheters, electronic design (both hardware and software), miniaturisation of electronic circuits and mechanical components.

4.2.1.4 Applications, Spin-Offs, Market Prospects and Evaluations

The potential technological repercussions of this research project in terrestrial fields of application are many. Just remember the Gore-Tex designed for the inside of the overalls for the Extra-Vehicular Activity (EVA) and now very popular throughout the sportswear. The experimentation of new fibres and fabrics for the clothing of astronauts that I led with Benetton Group has had immediate spin-offs in the sectors that concern sportswear (extreme, competitive, but also mass) and helped to elevate the comfort and performance of clothing, not only sportswear. Nowadays, it is normal to buy underwear and clothing which integrate thermoregulating systems and antibacterial properties, but at the beginning of 2000s, it was an extraordinary innovation.

The research and application of new chemical and physical micro-sensors (Sorin Biomedica Cardio), but also the development of monitoring systems have had technological repercussions in the research activities of cardiac micro-sensors and external tools (programmers) that have been used to receive and send signals and commands to implantable prostheses.

4.2.1.5 Preliminary Development Plan of the Entire Project

The project development plan, as required, has been divided into a *feasibility study* and an *implementation phase*. While the f*easibility study* was entrusted with the phase of *consolidation of the concept* and *identification of detail* of what to try on board the ISS, obviously avoiding to carry out verifiable experiments also on the ground, the next part was entrusted with the task of implementing and producing in detail what would have been defined during the *feasibility study*. Then, at the end of the *feasibility study* and on the basis of the *detailed proposal of experiment* defined in that interior, the *implementation phase* would have been developed on two parallel, but well distinct lines, the *experiment development* and the *design development* of the *Integrated Clothing System VEST* that allowed its implementation.

In this approach (already followed by Alenia Aerospazio Division Space during the EUROMIR '95 mission), the two parallel lines of development guarantee a dialectic between *who designs the experiment*, and therefore requires certain equipment, and *who designs the equipment*, and therefore provides the tools for the feasibility. This approach was defined during the feasibility study with all the partners already involved. The *pre-mission development* part was evaluated in a period of about 14–16 months and preceded the *mission* of about 6 months (as it is thought to be the time required in the programme for the ISS), an interval where the activity is reduced to small adjustments. This is a conventional choice due to the impossibility to know in detail how the equipment and the times connected to these procedures will be launched and installed on board.

The *design development* of the *Integrated Clothing System VEST* was foreseen achievable within one year, including component verification and acceptance. For the launch and the integration on board of the ISS, for which it is thought possible the use of a part of a standard rack, it has been used the knowledge of Alenia Aerospace Division Space, and especially of the programme European Drawer Rack (EDR) developed on contract by ESA.

Collections of astronauts' data were foreseen, both before and after the mission, for checks of comfort and fit. During the *development phase* of the experimental protocols, and the biomedical research group discussed and obtained the necessary permissions to operate with human subjects, although no data collection technique used was intrusive. It was expected that the activity on board within a mission (3 or 6 months), even if once tested the efficiency and comfort of the clothes, nothing prevented the crew to continue to use them. It was also thought possible to carry out the experimental part during the normal course of the various activities, using for some periods to be defined by the clothing proposed by us instead of the traditional one used on board.

In order to demonstrate the best efficiency of the proposed *Integrated Clothing System VEST*, a programme dedicated to each of the two principal parts of the capsule collection (work clothes and clothes for physical activity) would have been appropriate for one or two astronauts within the space mission.

4.2.2 Plan of Activities for the Feasibility Study

The *feasibility study* phase was foreseen for a duration of 10 months and begun with a detailed *analysis of the mission scenarios* of the ISS. This part was divided into three sub-parts, namely

1. the *analysis of the scenarios proper*, which would allow us to identify where, how and when to enter the experimental phase on board, what to provide and how to provide the necessary equipment and also what time and documents were required for access to the on board phase;
2. the *analysis of the requirements* to be met by the on board equipment and those imposed by possible transporters (see EDR or MPLM);
3. the *analysis of the scientific requirements* that would allow us to correctly validate both the concept and the equipment.

After the analysis of the mission scenarios we started the development of the *concept design* that was divided into two sub-parts: the *integrated concept* that identified the various components and the *detail of the various components*, both specific to the experiment and necessary to support it, considering the interfaces with the transport system, and then with the International Space Station (ISS). When the definition of the equipment had reached a stable stage, it would have been possible to define how to perform its *validation* in more detail; then the *experimental part* would have been developed (this activity was also intended to involve external experts in order to ensure full scientific correctness).

In the *experimental part,* we would also define which data should have been used for the validation of the *concept design*, in order to allow the completion of the definition of the *Integrated Clothing System VEST* with data collection systems. In fact, the last phase of the *feasibility study* proposed to define, on the one hand, *which materials and which data collection techniques to use*—also identifying if and what technological developments would have been necessary—and on the other hand, to prepare the final proposal for the actual testing phase. The proposal for the *experimental part* would have been dedicated to identify, with the support of the ASI, the possible points in common with other proposals to see if there were possibilities of optimization between experiments, or if we could have expanded the use of the information collected. This because it was believed that the development of the proposed equipment could have also been useful to disciplines of the human sciences. At the end of the *feasibility study,* it was expected to be a report of the work carried out both in the form of a document and as a presentation to the ASI. Especially at the beginning of the feasibility study, they were foreseen as a series of working meetings between the various partners, although only three of the formal meetings have been planned, namely, the *start-up meeting*, the *consolidation meeting* and the *final presentation of the results.*

The plan foresaw also a complete project regarding the organization of the *works packages* to describe the organizational structures of the complete project and that of the feasibility study. From this, we could see how the two parts were organized and who were the various responsibilities, even if the *multidisciplinary group* of the project acted as much as possible as an integrated group having to share very different experiences and know-how.

4.2.2.1 Design Goals of the Integrated Clothing System VEST

The core of the *feasibility study* was the development of the concept of a collection of garments specific for Space, that I personally created and developed, and of which I report here the results entering in a more detailed description of objectives and design hypotheses from the Design point of view. The new clothing system is designed for life and work on board the ISS so as to increase the comfort of the crew through the search for new fabrics and clothing specific to the confined environment and microgravity. It also provides integration of sensors in the clothes, for the monitoring of biological parameters, and prostheses, to facilitate the movements, and consequently, the Intra-Vehicular Activities (IVA) of the astronauts.

The principal aim of this project is to identify the main requirements of fabrics and clothing to be used on board the International Space Station to develop a new integrated clothing system able to enhance the body and make it more comfortable and efficient crew activities.

The state of the art does not provide for any type of specific clothing that takes into account the many inconveniences caused by living in a confined environment and in conditions of microgravity: significant variations in internal temperature, biological and postural alterations, lack of privacy, the need to do many hours of exercise every day to maintain muscle mass and tone and to counter bone decalcification, resulting in increased perspiration, are just some of the problems astronauts face *without having adequate and diversified outfits* to carry out the various activities comfortably. In addition, equally important is the *pleasantness* that comes from wearing a dress that does not necessarily resemble a military uniform, but rather that enhances the personality and identity of the astronaut. Having a good sense of self means *feeling comfortable with your body and your second skin*, to the benefit of an overall improvement in on board social relations which inevitably result in greater *operational and performance efficiency* (Dominoni 2003a, b).

The study of fabrics, cuts, shapes, colours and accessories to diversify the garments and provide a wider choice for astronauts is essential to increase the level of well-being and improve the quality of life and social relations on board. A *second skin* could satisfy the functional needs of the body and spirit.

The objectives of the project are, therefore, aimed at increasing the degree of well-being and efficiency of the astronauts by improving their psychophysical condition through the *wearability* of the garments, the *aesthetics*, thermal stability and body hygiene, and the monitoring of the biological and physiological vital functions of the crew, thanks to the use of sensors integrated in the garments without being invasive. The project provides for the possibility of prefiguring different types of clothing for the various activities on board, such as work, rest, physical activity and underwear. The design of the clothes also considers the problems related to the *storage* of the material on board, with particular attention to the *packaging* and the minimum size of the same, and provides treatments that facilitate maintenance and possible *washing* (for the time being not yet resolved, since the clothes are packed and brought back to Earth dirty). For this specific problem I was exploring new possibilities, as a system of washing the clothing to be integrated onboard the ISS, in which washing technologies[2] were closely related to the chemical characteristics of the fibres and detergents that would have been identified during the project.

[2]At the same time, I was designing a system to wash clothes in Space without water, using vacuum, in which I involved Electrolux Zanussi in a Pre-Feasibility Study along with ASI, but then, unfortunately, the lack of funds did not allow the project to get to prototyping and verification on board the International Space Station (ISS).

4.2.2.2 Project Researches and Hypothesis

1. **Increase the Level of Comfort and Efficiency**

The absence of gravity is generally considered an advantage for movement in Space. Once accustomed to microgravity, the body moves with minimal efforts and is able to perform *acrobatic maneuvers* with extreme ease. However, in order to exert forces, the body must be anchored so that the opposing force can be developed for the active exercise of the force-vector. One problem that occurs widely during the first few days in the absence of gravity is the decrease in motor performance.

The astronaut finds it difficult to accurately estimate the amount of physical work required for certain operations and also takes longer to perform the various activities due to the lack of landmarks and strength based on gravity. After a few days in microgravity, the movements and activities of the astronauts become more precise and the perception of the difficulty level decreases. When the crew returns to Earth, after a long stay on board, the vertical stability decreases along with the reduction of muscle strength. It should also be noted that astronauts tend to drop things for the first few days, believing that they can float in the air as in microgravity. These problems disappear after about a week of stay. It is not yet predicted how long it will take to recover the effects of a 1/3G environment like Mars.

Microgravity has a decisive influence, as well as on the human being, on the strength-weight of objects, increasing the difficulty in use and control. The use of Velcro® is in fact the only current efficient and less invasive system to fix objects to the structure of the station or to the body of astronauts, mainly thanks to the ease of use and the minimum size, in terms of volume and weight. The increase in comfort, to the advantage of greater well-being of the crew, directly affects the operational efficiency and performance in carrying out the activities on board.

The *Integrated Clothing System VEST* aims to facilitate the movement and physical and operational activities of astronauts through a series of equipment and prosthetics, incorporated in clothing, that allow the anchorage of the crew to the internal structure or parts of it and the transport of tools and objects through their Velcro® fastening on the clothes worn (Fig. 4.1).

2. **Provide New Parameters of Wearability**

Currently, astronauts wear clothes that they bring to Earth and that are not suitable for the particular environmental conditions on board. In Space, there is a substantial change in the posture and size of the human body. If relaxed, the body assumes a semi-harvested position, in which the angles of the knees and elbows are about 130 degrees, changes the pelvic angle and flattens the curvature of the spine in the lumbar and thoracic section, giving rise to a body extension of over 10 cm. The head and spine bend forward and the upper limbs fly upwards to the trunk 450°. The new Neutral Body Posture (NBP) to which crew members are subject in microgravity conditions is similar to that taken by the body underwater and seems to result mainly from a new balance of muscle forces in relation to the tension of the

Fig. 4.1 A rendering made for the concept design of the *Integrated Clothing System VEST* created by the author shows removable textile pockets to transport various items and tools through their Velcro® fastening on the clothes worn. Credits by the author

tissues acting on the various joints. While the NBP in Space does not in itself have contraindications, apart from the inconveniences that can occur by bending forward or trying to sit or stand upright as on Earth, it is necessary to carefully consider the new implications between operator and instrument.

The design of a space clothing system must take into account in an essential way the Neutral Body Posture in Space and consequently provide new parameters of fit, weight and fall of the fabric, cuts and seams of clothes, adhesions and elasticity in relation to the postures and movements that are created in the absence of gravity. Adhesions should be carefully considered at the joints and ends to prevent excess tissue from becoming entangled in structural parts within the ISS.

3. **Consider Aesthetics as Fundamental**

The study of fabrics, shapes and colours and accessories, to diversify the garments and provide a wider choice for astronauts, is essential to increase the level of well-being and improve the quality of life and social relations on board. The pleasantness that comes from wearing a garment that is able to enhance the identity of each astronaut is a very significant parameter if we think of the condition of confinement to which the mixed and multi-ethnic crew is subjected. The *Integrated Clothing System VEST* should be able to differentiate the various specificities of astronauts and backgrounds while considering the new wearability parameters.

Fig. 4.2 An image of the transposition of the concept design *Integrated Clothing System VEST* created by the author into a collection of clothing worn by the astronaut Roberto Vittori during an experiment on board the International Space Station. Credits by ASI

During the second technical meeting, a particular need emerged for astronauts to wear garments in orbit that do not differ much from those used during training on Earth, to facilitate the adaptation of the crew on board. So sci-fi style is not recommended because the crew prefer very normal clothing. What a pity for me!

Recreating a home and family environment seems to be the most useful and attentive design choice to the requests of the users who will use the clothing system living and working on the ISS.

The *Integrated Clothing System VEST* can then consider the opportunity to design garments for both ground activities and those in orbit, differentiating the wearability, and then the cuts and fall of the fabrics, in relation to the different environments, but maintaining unchanged comfort, aesthetics, with the same types, variations in colours and accessories, and thermoregulatory treatments, breathable and antibacterial on the fabrics (Fig. 4.2).

4. **Ensure Thermal Stability and Body Hygiene**

The temperature inside the ISS, under normal conditions, has a variation between 18.3 and 26.6°, while in emergency conditions, the variation expands between 15.6 and 29.4°. It is, therefore, important to be able to wear garments that are able to counteract external thermal changes, while maintaining constant body temperature. The confined environment and the scarcity of renewable resources, such as water and air quality, pose problems that inevitably affect the hygiene conditions of the crew. Limiting the contraindications due to excessive sweating, especially during physical activity, becomes a considerable advantage that is extremely effective when applied on the second skin. The use of thermoregulatory fibres and fabrics, which absorb excess heat and then release it when the body cools, breathable, were able to limit the harmful effects of perspiration by transferring excess sweating outside the garment, and were antibacterial, thanks to silver particle finishing, which can help to improve the thermal stability and hygiene of the astronauts on board. In addition, tissues and garments should help to reduce skin flaking by retaining particles.

5. **Allow the Control of Vital, Biological and Physiological Crew's Functions**

Monitoring of body values such as temperature, heart rate, circulation, breathing, physical stress, Fatigue, etc., is currently carried out with equipment that requires long application times of sensors and limits astronauts in movements and in the performance of activities. One area of the design of the *Integrated Clothing System VEST* involves the inclusion of sensors in garments that will allow the detection of astronauts' bio-signals without interfering neither with the subject nor by its activity, and without requiring a direct and conscious application. An essential condition is that the sensors are in direct contact with the body, or in any case inserted into tissues adhering to the body.

4.2.2.3 Concept Design Development

The concept of the *Integrated Clothing System VEST* provided different types to meet the requirements (identified by the technical note NT1) and the study, included in this activity, for the monitoring of biological parameters (technical note NT3). The need to position the sensors in direct contact with the skin, combined with the need for greater comfort that facilitates the activities of the crew, has directed the design choice towards the creation of a specific underwear that is able to integrate the sensors, while giving the astronaut complete freedom. The underwear consisted of two pieces, differentiated for men (t-shirts and shorts) and women (brassiere and briefs), and allowed the monitoring of biological parameters. The garments packaging also uses technological processes that allow to obtain a seamless tubular mesh, for greater comfort and a more snug fit. The other types of clothing include specific garments for the three most important activities and incidences with respect to the allocation of time identified in the technical note NT1:

- the work
- the rest
- the physical exercise

All types of the *Integrated Clothing System VEST* were foreseen made of antibacterial, anti-odour, thermoregulatory, breathable and antistatic fabrics, in different percentages depending on the specific characteristics of each garment.

The Work Clothes

(t-shirts, long pants, short pants)

Privilege the wearability, in relation to the environmental conditions dictated by microgravity, and the comfort and consequent operational efficiency, to facilitate the movements and activities of the crew.

The garments are equipped with tools (slings) that are able to anchor the astronauts to structural parts (rail-straints) and between them; they provide Velcro® inserts to attach tools to the body during routine maintenance operations to be carried out on board, and integrate small light sources to facilitate various activities, which often take place in very narrow and poorly lit areas of the ISS.

The Sleeping Clothes

(t-shirts, long pants, short pants)

Privilege comfort, to provide more freedom of movement to astronauts during sleep, but also entertainment services.

The garments also integrate small light sources to allow you to read or send emails; they provide the ability to listen to music through speakers fixed on the garments at the shoulders.

Clothes for Physical Exercise

(t-shirts, short pants)

Privilege comfort, thermal stability and body hygiene to counter excess sweating during movement.

The tissues of the garments for physical exercise are foreseen, therefore, to have a higher percentage of antibacterial, anti-odour and thermoregulatory treatments compared to other types of the collection.

4.2.3 VEST Clothing Support System on Orbit Validation

The experiment was scheduled during the Italian *Marco Polo* space mission in April 2002 about which I was the Principal Investigator. Aim of the experiment was to demonstrate that the provision of a new *VEST Clothing Support System* to the astronauts would increase their well-being, would allow to collect useful data with less impact on the crew schedule and to prove that in general with *VEST* equipment less mass and less volume are required for the clothing system. The results of *VEST* experiment (Dominoni 2005) have allowed to obtain specific requirements for a specific clothing system in microgravity and confined environments that had not been proposed by any research group before. At this regard, it follows a brief description of the garments used by the astronauts on board the Shuttle and the International Space Station (ISS) with technical requirements.

4.2.3.1 Garments Used in Human Space Missions

During human space flight the crew members worked inside the Space Shuttle in shirt-sleeve comfort. Prior to a mission was outfitteing from a selection of clothing including flight suits, trousers, lined zipped jackets, knit shirts, sleep shorts, soft slippers and underwear. Today things are more or less the same. Nothing is changed. The crew wear the same clothing that they use on Earth without specific requirements for the space environment. There are only some characteristic with small adaptation of the garments to microgravity such as closable pockets for storing some objects as pens, pencils, data books, sunglasses, a multipurpose swiss army knife, scissors or a flashlight as well as Velcro® attached to the shorts and trousers where a Nomex pocket attaches to carry their items. The materials are supposed to be *flame retardant* even if not all the clothing because underwear, shirts and polo or rugby type shirts are made of cotton 100%. Shorts, trousers and flight jackets are made of Nomex, which is a rude fibres very *unconfortable*. For their feet they either wear 100% cotton socks, slipper socks made of wool and leather or Polartec fleece socks. Because the lack of space and the necessity to limit weight, or a 7-day flight, a crew member had one shirt per day, 2 or 3 pairs of socks/slippers, 2 pairs of shorts and 1 or 2 pairs of trousers, 1 sweater or jacket.

With the International Space Station (ISS) NASA could collect some crew's preferences from the experiences of the previous missions: both shirts (long sleeve and short sleeve) are off-the-shelf clothing items and must be made of 100% cotton. Long and short sleeve shirts are selected and then procured from a vendor. They are limited to a few styles to select from. The colour and size vary according to each choice. They may have monogramming, a mission patch, or NASA logo sewn on. They are available in sizes small for females to through extra-large for males.

The *incremental innovations* compared to the Space Shuttle was first to try to make *more comfortable* the clothing in Nomex with finishing on cotton fibres to make the cotton fire retardant and allow astronauts to wear more soft flight jackets, trousers and shorts. The original cotton twill material was treated to enable the assembly to comply with flammability requirements in a 30% oxygen-enriched atmosphere. The flight jacket was proposed royal blue colour, with multiple pockets, featuring a Nomex knit waistband and cuffs. It ranges in size from small/regular (chest/sleeve) to extra-large/long, and maybe worn with or without the jacket liner. The crew preference flight jacket liner was a natural coloured vest made from flame-retardant treated cotton cloth that zips-out from the jacket assembly. The line exterior had a layer of Nomex batting quilted to it for insulation.

The second suggestion by astronauts was on *wearability* to have a wide range of sizes to accommodate either male or female, and flight trousers feature an adjustable waistband, two side and hip pockets, a zippered pouch on the left calf, and several strips of Velcro® across the legs to allow for mating with the removable pocket. Featuring an elastic and drawstring waistband, the crew preference was flight sleep shorts which had two side pockets, one flap hip pocket, Velcro® across each leg for pocket mating and a white liner.

On ISS were introduced also *synthetic materials* in clothing as the *Lightweight Kit* consisting of a zippered shirt and a pair of shorts. The shirt was made of thick cotton. The shorts were made of polyester and feature a zippered fly, elastic waist and two zippered pockets. Lightweight clothing colours are crew's preference with some limitations. This clothing kit is usually worn during periods of interaction. Such as crew exchange, TV interviews and casual times. Another example is the *Operator Suit*, a sleeveless, multi-pocketed and zippered coverall with stirrups. It is composed of an antistatic, synthetic outer layer with 15% Lycra, paired with a cotton lining. The Lycra in this garment provides a *feeling offorces* being applied to the body. The suit features two zippered chest, thigh and rear pockets, along with a zippered left calf pocket. This coverall comes in a variety of colours, which are crew's preference, and 2–4 sizes (accommodates 24 sizes).

4.2.3.2 Fibres and Fabrics Innovations of VEST Project

The *Clothing Support System VEST* introduces important innovations which increase the well-being during Intra-Vehicular Activities (EVA) as the use of thermoregulating fibres and fabrics, which absorb excess heat and release it when body temperature falls and are capable of curbing the noxious effects of perspiration by transferring excess sweat to the outside, and enables one to fight bacterial proliferation and body odour, thanks also to a special small silver particle finish. All this contributes to improving body thermal stability and overall hygienic conditions on board. The basic principle informing the search for materials, production procedures and crucial technologies for the realization of the new items of clothing—as developed in the course of the project—was that any item of clothing or accessory is to be considered a multi-composite product, which stems from the interaction of a number of technological parameters. The research of fibres and fabrics was carried out in four main areas of technology:

1. *yarn technologies*: yarns with phase-changing microcapsules for active control of body temperature, silver-coated yarns so as to control germs and fungi due to perspiration and elimination of ensuing any body odour, and yarns containing micro-particles of organic compounds so as to kill off bacteria and fungi due to perspiration and ensuing body odours;
2. *weaving technologies*: orthogonal weaving is carried out with shuttle looms with yarn weft and warp orientation, respectively, at 0° and 90°, circular weaving is carried out with circular looms for the creation of tubular knitwear;
3. *fabrics technologies*: considering that one of the primary goals in the *feasibility study* of *VEST* was to choose garments capable of reducing paying load at take-off as compared to standard paying loads, the researches I led have

Fig. 4.3 Among which the innovations of the *Integrated Clothing System VEST* designed by the author, great attention is given to the garments' wearability of the whole collection, that follow new parameters tailored on the Neutral Body Posture (NBP). Credits by ASI

highlighted how, in relation to the 100% cotton fabrics currently used in a number of garments, mixes of cotton and acrylic fibres allow for a substantial reduction in the mass of the garments (provided one compares two items of clothing made the same way). By way of example, a 40% Cotton and 60% Thermoregulating Outlast Acrylic would allow for the creation of a fabric weighing at least 15%;

4. *treatment technologies*: compatibility between the fabrics treated and the user's skin owing to the peculiar conditions on board. The research of the best technology was also addressed to the garments manufacturing with a special attention being devoted to accessory design and the choice of the most suitable materials as well as to the packaging which foresaw vacuum-packed, and the best solutions adopted in space flights.

4.2.3.3 Tailoring and Wearability Innovations of VEST Project

As far as aesthetics is concerned, the clothing wearability for Space environment is built on the Neutral Body Posture (NBP) which the crew members adopt in conditions of microgravity and is similar to that adopted by the body underwater and seems to result above all from a newly found balance between muscular and the tension of fabrics acting on the different joints. During the project research, I

Fig. 4.4 In this picture it is easy to observe the gap between the wearability of the *Integrated Clothing System VEST* designed by the author, and the normal clothes used both on Earth and in Space without to consider the effects of microgravity. Credits by NASA

discovered that this particular posture is very similar to that adopted by snowboarders—with limbs bend, knees and elbows upwards and head bends forward—and therefore, some functional details as cuts and seaming designed to adhere to the body of the snowboarder for a better wearability have been directly applied to the Clothing Support System VEST. The whole clothing system VEST is based on the concept of flexibility so as to answer the project requirements identified, and notably the two that most influence the design and style of clothing: comfort—and, as a result, operative efficiency—and wearability, and therefore aesthetics. The integrated clothing system foresees specificities and solutions ensuring that, once the garments have been put on, the adherence and elasticity of sleeves and trouser legs and of the waist may be adjusted, thanks to strings threaded through the single items of clothing, which may be widened or tightened up and secured with Velcro®. This makes it possible to adapt single garments to the changes the body undergoes in conditions of microgravity, be they physiological (fluids tend to shift to the upper part of the body, causing it to swell), morphological (atrophy of the lower limbs and lengthening of the spine) or postural (when in orbit one assumes a neutral semi-curled position). Moreover, the possibility to modify adherence and elasticity in relation to the 1G environment enables the astronauts to wear the same kind of clothing also on Earth, during training and in the preparatory stages of human space missions (Fig. 4.3).

4.2.3.4 Aesthetics Reflections of VEST Project

When I was asked to design a collection of clothing for astronauts, my first wish was to create clothes with a strong identity that would make people immediately understand that the person who wore them was an astronaut who was living on board the International Space Station (ISS). Such an extraordinary role, in my opinion, required garments with a strong aesthetic value, which, while not compromising comfort and functionality, were similar to superhero costumes that recalled the collective imagination of sci-fi and future scenarios. But during the various meetings of the *feasibility study* and then for the preparation of the experiment and, above all, the briefings with the astronauts, I realized that there was a certain resistance, mainly due to the fact that having to live in an environment already very different from the terrestrial one they wanted certainties that made them feel at ease, at home. Contrary to my expectations, as regards the formal and aesthetic choices of the *Integrated Clothing System VEST*, astronauts feel the need when in orbit to wear cloche not all that dissimilar from those used during training on Earth, so as to facilitate their adjustment to life on board. As already underlined previously, to recreate a domestic and family environment seems to be the best approach and the one best answering the requirements of the astronauts wearing the new *Integrated Clothing System VEST*. Banal garments for them was a way of feeling normal, very normal. I should have looked for innovations in other fields, in the search for materials and fabrics, in the wearability of garments, and in their adaptability according to the various activities to be carried out on board. I was, however, intrigued by the idea of making some reflections on the aesthetics in Space and on the value of wearing clothes in such extreme living situations, which concern not only the effects of microgravity, but the fact of living for a long time in a restricted and confined environment, without the possibility of leaving, and where they create forcibly group dynamics in which, I think, it is important to affirm their identity while they maintain a strong group spirit (Dominoni 2003a, b).

The choice to explore the behavioural dynamics with regard to aesthetics aspects in an extreme environment as Space allows me to get into focus the human problems and the psychological and social implications that could derive from it. The conditions of constant emergency of the crew's members on board the ISS, the confined environmental and the microgravity effects cause heavy stress that made astronauts more sensitive and vulnerable: the interpersonal relationships are altered and the harmony of social sphere—the cultural and ethnic peculiarities of the different crew members—is seriously compromised. In this context, each behaviour is increased, and the actions are laden with meanings that on Earth could be of secondary importance. The aesthetics in microgravity doesn't escape from this principle: the extreme conditions in which astronauts live require a particular attention to aesthetic values to encourage a high level of personal satisfaction, and therefore a more quiet acceptance of one's self. It has to be said that if in common earthly conditions to have a positive perception of one's self means feeling at ease with one's body and its second skin—which leads to an overall improvement in such social relations, and therefore, of increased efficiency in terms of both operation and performance—in Space; the aesthetics values are essential for the well-being of each individual and have to be considered fundamental for the success of the mission. On board the ISS the performance equipment must be extreme, the clothing inclusive, and it is necessary to study a high technological level of materials and production's processes compared to products used on Earth. This means that the clothing worn by astronauts should improve the aesthetics and comfort through innovative technological and suitable choices to take advantage of the environment (Fig. 4.4). To study in depth the aesthetics meaning of the crew's clothing could provide interesting research tools with application in different contexts as technological solutions selected for spatial environment that could foresee important spin-offs on Earth, from extreme to domestic contexts. It should be borne in mind that the study of life in Space is still in its early stages and that Space itself is after all a relatively new environment for human beings. Its effects have been recorded by collating the experiences of a few people, most of whom have spent but a few days in Space, while just a few have spent several months in Space.

Where do designers draw inspiration from when creating styles for Space closing totally unrelated to the unimaginative one—a far cry from the kind of clothing described in books and science fiction films—used during space missions? What does the industrial designer deem to be of paramount importance when setting out creating a new and aesthetically pleasing *Integrated Clothing System VEST*? The solution consists in striking a balance between functional needs—related to comfort and the countermeasures called for by the microgravity environment—and subjective needs—which include personal taste and the pleasure to be derived from wearing a given item of clothing (Dominoni 2003a, b). If we then wish to analysze the issues of style and aesthetics, like most users, astronauts do not seem to have any clear ideas. It is instead the task of industrial designers to come up with a number of solutions, and just as many variables, by giving shape to the needs expressed by the users, but also by embodying potential ones in real and usable products. The study of fabrics, of shapes, of styles, of accessories and of colours aimed at the diversification of garments, and at providing astronauts with a wider choice is essential if one is to improve the level of well-being, the quality of life and social relations on board. Just as important is the pleasure that is to be derived from wearing clothes that do not necessarily smack of military uniforms but to enhance instead the personality and identity of the astronauts. This parameter becomes very important if we consider the confined environment the mixed-sex multi-ethnic crews have to live in. To increase the degree of well-being and efficiency of the astronauts, to ensure garment wearability, bodily thermal stability, hygiene and the monitoring of biologic and physiologic functions are the project requirements adopted to counter the adverse effects in Space and to improve living conditions within confined spaces, which have to be shared by a number of individuals living under mental and physical stress. Such project requirements fully respect NASA indications as to fabric composition, weight, volume and storing, and therefore, greatly influence overall project aesthetics.

4.2.3.5 The Experiment Sessions

VEST experiment was specific for the *Marco Polo* mission of the Italian astronaut Roberto Vittori, scheduled in April 2002 on the Soyuz Taxi Flight. The scope was to validate the on-orbit functionality and concept innovation of *VEST Clothing Support System*, consisting of a complete garment set. Aim of the experiment was to demonstrate that the provision of a new integrated system of garments to the crew would increase their well-being, which would allow to collect useful data with less impact on the crew schedule; and to prove that in general with *VEST* equipment, less mass and less volume are required for the clothing system. Diversified items of clothing, more than the underwear garments included, foresaw the three most important and most significant activities—in terms of time management—as identified in the mission analysis: work, night rest, physical exercise. The requirements foreseen for the crew are very simple: to wear *VEST* clothes and to provide opinions concerning their psychological and physiological well-being, therefore, to provide comfort, garment wearability, aesthetics and efficiency of thermal stability and bodily hygiene on board. I used specific questionnaires. No pre-flight training was necessary for the *VEST* experiment to be put in practice. Astronaut Roberto Vittori was informed about the garments set to wear for each session, about timing and procedures to follow on board. He was aware of the experiment purpose and he was willing to give his opinion answering to the questionnaires.

The experiment sessions, apart the first one dedicated to physical exercise, was extended during normal work and sleeping activities on board the ISS. For the other two session types, the astronaut wore *VEST* provided clothes while performing mission planned activities. In addition, also the time devoted to remove clothes from *VEST* bag, astronaut dressing, astronaut cloche removal, clothes re-stowage and questionnaire filling in was considered as astronaut time allocated to the *VEST* experiment. During work and physical exercise sessions, *VEST* experiment required also the part-time involvement of another cosmonaut for taking pictures and video recording.

Data have been collected and analyzed by questionnaires, video and still images. Fundamental topics as *usability* and crew *well-being* have been checked taking into account the astronaut's opinions wearing the *Integrated Clothing System VEST* during the various activities on board. Body postures, movements and clothes *wearability* in microgravity were the principal parameters for questionnaires, photos and video recording. The four questionnaires—one for each three activities plus underwear described before and one testing the wearing easiness—was based on questions with multiple answers. Questionnaires submitted to astronaut's opinion included also to consider reports of negative impressions. Several questions foresaw a comparative analysis between *VEST* clothing system and usual garments on board with the aim to verify the *efficiency* of the new garments.

VEST Clothing Support System On-Orbit Validation's experiment has been based exclusively on Roberto Vittori experience, and therefore his opinion is not representative of any statistical result, as are most of the experiments conducted on the ISS. Even so, the experimentation results were extremely significant for the validation of *VEST* clothing considering the study's development in the future. Vittori's opinion was important in order to help garments design concept for a longer and more complex mission where several crew members of different gender, race and size could have worn a new designed set of *VEST* clothing. I am grateful to Roberto Vittori who shared with me the interest and the enthusiasm for the innovative garments during the experiment activities. The results of the questionnaires demonstrate the full success of the experiment and the importance of having clothes dedicated to Intra-Vehicular Activities (IVA) that make more comfortable and pleasant living and working in Space. Thanks to *VEST* project, Vittori has known the Design meaning and appreciated the ability to shape the needs of astronauts following their needs and finding solutions more favourable to the confined environment and microgravity.

My role of designer, in this case, consisted in mediating the needs of the user-astronaut with the peculiarities of the space environment, and consequently, in considering the opportunity of designing garments both for ground and in-orbit activities, by differentiating wearability, and therefore tailoring and fabric drop in relation to the different environments, but without renouncing comfort, and aesthetics, or favourite typologies, shades of colour and accessories, and the same thermoregulating, anti-perspiration, antibacterial, anti-odour and antistatic treatment for fabrics. At the end, the task of Design is to translate project requirements into opportunities that can be expressed through aesthetic values.

The proof of the success of the experiment VEST was the request that was made to me personally by Roberto Vittori to continue in the implementation of the project providing other garment sets for his subsequent *Eneide Mission*, scheduled in April 2005, on the Soyuz Taxi Flight: with the experiment *GOAL* I had the chance to affirm again the value of new requirements and items to make comfortable and more efficiently his on board activities during human space flight.

4.2.3.6 Meaning and Spin-Offs of VEST Project

The development and completion of the *VEST* research project and experiment open up extremely interesting opportunities of generating—both in the short and in the long term—meaningful technological spin-offs both for the clothing textile industry and for other fields of production. In recent years, the textile sector for both leisurewear and sportswear has been undergoing meaningful changes and companies have increasingly been focusing their attention on innovative fabrics and technologies. *VEST* introduced new product technologies which address the issues of the user's comfort, well-being and safety as well as ensuring energy saving and being environmentally friendly. As a rule, new technologies have higher costs than already consolidated ones, especially when first introduced onto the market, and the support of a high profile scientific activity, such as that developed in Space projects, capable of ensuring the quality and performance of the new product, which cannot but be beneficial for the promotion of that self-same technology with the end-user.

For example, the *Integrated Clothing System VEST* for confined environments and conditions of microgravity—which foresees the use of antibacterial and thermoregulating treatment for fabrics capable of counteracting the former the effects of body perspiration, the latter the variations in temperature of the ISS—might prove to be extremely effective also here on Earth in the course of sports competitions and, by dilating the sphere of application, also in the course of any given individual's normal physical exercise. Spin-offs will be recorded in the use of fabrics specifically developed for use in Space but also in the solutions of increased comfort and flexibility adopted for casual and sportswear. Apart from the far-reaching effects to be recorded in the field of textiles for clothing, one can easily picture the spin-offs to be had in the field of furnishing fabrics for the automotive industry, where the functional problems related to active thermal regulation and the killing of bacteria can be just as important. When the *feasibility study* of *VEST* started, in the years 2000, the use of antibacterial and thermoregulating fabrics was a remarkable innovation. The spin-offs I foresaw have shown that these technologies have extended to many areas of applications and productions, and now many garments, from underwear to competitive sports, use these fabrics and textiles that were born for our project in Space, as basic requirements for active textiles and clothing.

4.3 Couture in Orbit. A Capsule Collection for ESA

I believe the project *Couture in Orbit* is the first important step to communicate to a huge public the role of the European Space Agency (ESA) in developing innovation and inspiring people through the design and fashion language, that celebrates space exploration and research exploitation and their impact in many fields of our life. Thanks to this initiative, ESA has experimented the value of Space Design in finding disruptive solutions to apply in many fields on Earth to increase the well-being of our society looking at Space as a booster of creativity.

The idea of *Couture in Orbit* was aimed to mark the missions of five European astronauts—from Italy, the UK, France, Denmark and Germany—who flew on board the International Space Station (ISS) between 2014 and 2016 organizing an innovative project of a *capsule collection* which was presented during an event on 25 May 2016, at the London Science Museum, involving five fashion design school from each of the astronaut's home countries.[3] I was invited by ESA to lead the Italian side of *Couture in Orbit ESA POLIMI* involving my students of the School of Design, Master of Science in Design for the Fashion System, at Politecnico di

[3]*Couture in Orbit*, ESA, Taking Technology and Fashion to Higher Levels, 25 May 2016, https://couture-in-orbit.tumblr.com/ (accessed 10.09.18).

Milano, with the aim to create a collection of garments inspired by space scenario, using ESA technologies, which were designed to be worn every day on Earth.[4]

Working with *Couture in Orbit* was a great challenge for us to explore fashion in the age of technology and understand better how we can be more conscious of the transformation technology brings. I was very excited because I believe space innovations will have a strong influence on how people behave and perform in the next future.

Before to go into the description of this case study, I think it is necessary a short excursus on the interaction of Space and the fashion design industry before the project *Couture in Orbit* to demonstrate the strong link between Fashion Design and Space, starting with *Space Age*, which has redesigned a future creating a new style inspired by colours, shapes and space materials, and which evolving in technological research integrated into garments, led to the development of intelligent fabrics, smart textiles and wearables that are part of our daily life.

4.3.1 Space and the Fashion Industry

Space and fashion industries often influence each other with spin-offs and spin-ins. The most famous spin-in, from Earth to Space, is surely the use of Velcro® by NASA's astronauts during their human space missions to fix objects easily in microgravity: it is a patent registered in 1958 thanks to the intuition of the swiss engineer George De Mestral who, observing the alpine thistle properties consisting in micro elastic hooks, tried to apply the same principles to facilitate assembly operations creating Velcro® which became immediately very common in the clothing system all over the world. There are many examples of spin-offs, from Space to Earth: innovations that have become more than familiar, common objects of everyday life, such as Gore-Tex® of the astronauts' suits (De Monchaux 2011) that—thanks to the light weight, waterproof protection, breathability and cold resistance—has gone from Space applications to those of technical clothing and sportswear industry, becoming a basic component in thermal insulation.

One of the most fruitful and lasting collaboration with NASA is the industry DuPont, whose advanced materials and thermal protection systems have made possible human and unmanned space missions for nearly half a century. DuPont is the owner of brands such as Nomex® and Teflon®, the two fibres used in astronaut suits for their heat resistance performances, which guarantees security and protection to the fire brigade during fire impacts and first-aid operations. NASA developed recently Aerogel (the first molecules of Aerogel date back to 1931) to insulate space probes sent to Mars, and the first Earth application regarding fashion is Italian. Corpo Nove, a company specialized in technical clothing, used Aerogel to

[4] *Couture in Orbit*, ESA-POLIMI, European Space Agency & Politecnico di Milano make Fashion, 19 February 2016, https://couture-in-orbit-esa-polimi.tumblr.com/ (accessed 10.09.18).

design and develop the jacket *Absolute Zero*, that exploits the extraordinary thermal qualities of this material; it can insulate at temperatures between 50 and 3000 degrees Celsius, and with regard to its lightness, it is composed of 99% of air and is the lightest solid substance in the world.

With the support of ESA Technology Transfer Programme (TTP), Grado Zero Espace (the research and development laboratory of Corpo Nove) has developed another project using shape memory alloys to obtain Oricalco®: a textile having a content of Titanium of 45% and characterized by the extraordinary ability to recover any shape pre-programmed upon heating. With this fabric was made a shirt with long sleeves. The sleeves could be programmed to shorten immediately when the room temperature heats up. The shirt can be screwed up, pleated and creased, then, just by a flux of hot air, even a hairdryer, it can pop back automatically to its original former shape. ESA innovations were recently used in thermal underwear for the manufacturer Björn Borg as well as an Italian motorcycle-clothing manufacturer, Dainese, tailored ESA's Skinsuit to alleviate astronaut back problems.

Space exploration could become today a driver for growth and competitiveness involving industries and companies not strongly related to Space, as it has been its policy until now. The European Space Agency (ESA) is looking into novel ways to advance its strategic goals for space exploration that can include private sectors with the aim to create new products and services which are able to answer to the emerging needs of the society constantly changing and reach a wider audience. It is for that reason that, together with the traditional areas of interest for ESA consisting in user-driven exploitation of Low Earth Orbit (LEO) infrastructures, lunar and martian exploration, and research and development, we find, for the first time, the word inspiration: how could we consider this keyword's choice?

4.3.2 *Taking Inspiration by Space to Innovate*

How many times Space has inspired the human being? All the worlds of thought have looked at the universe: science, philosophy, religion, technology, literature. And also, all the disciplines related to art and practice: design, cinematography, communication, music and many more. Sky and stars continue to exert a strong fascination on every one of us beyond the ages and fashions. There are different kinds of innovation that can reflect Space: style innovation, meaning innovation or technology innovation, including spin-offs and spin-ins, transferring by Space to Earth and vice versa.

4.3.2.1 Innovation of Style

The fashion designers have been inspired by Space envisaging new scenarios explored by the first human space missions and the astronauts' experiences. The microgravity effect, with the possibility to fly and move without to feel weight, and the possibility to look at our planet from another and external point of view are the two most innovative and extraordinary conditions which are able to generate visual suggestions that can be translated into shapes, colours and materials: rounded volumes, optical white, and silver nuances as brilliant moon powder, mirroring metals and synthetic fibres.

The visions of the future of Pierre Cardin, André Courreges (Fig. 4.5), Paco Rabanne started during the 1960s with the Space Age (Topham 2003) which implied a great research on experimental fashion based on space styles white and lunar colours, and innovative materials, with the aim to foresee a new way to living and dressing capable of suggesting dreams and inspiring people. Paco Rabanne was at the helm of the Space Age fashion movement thanks to his use of unconventional materials, making clothes from plastics and metals. Indeed, his inaugural 1966 collection was titled *12 Unwearable Dresses in Contemporary Materials*, and his signature dresses comprising plastic discs or aluminium plates strung in a way that was reminiscent of chainmail. Paco Rabanne extended his research to popular science fiction when he designed the costumes for the classic cult *Barbarella* in 1968, which is set in the 41st century. As for today, under the direction of Julien Dossena, the fashion house has been returning to its futuristic roots with chainmail skirts, sheer Nylon® and Perspex® wedge boots. Many fashion designers have often turned to science fiction and its associated futurism for inspiration. The distinctive costumes from cult science fiction flicks have proved lasting influences, from Jean Paul Gaultier's looks in *The Fifth Element* to the punk-rock garb in *Blade Runner* and the robot from 1927 silent film *Metropolis*, which has been embraced by the likes of Karl Lagerfeld and Thierry Mugler (Dominoni and Quaquaro 2017a, b).

And this trend is continuing until now. *Star Wars* saga is the inspiration of the new fashion show for men of Versace in 2016: tracksuit in optical fibres, shirts with luminescent piping as in astronaut suits, outwear with embroideries like space debris, leather jackets decorated with crystals, jeans with imprinted entire constellations, and sneaker with lunar landing sole. The most surprising action inspired by space scenario was performed during the fall 2017 show by Karl Lagerfeld who organized a catwalk for a Channel at the Grand Palais in Paris simulating a C-branded spacecraft launch in the final part of the catwalk dominated by silver glitter garments and accessories as space helmets worn by the models. It was a powerful way of updating the space style in our contemporary world which looks at space tourism as a close and feasible opportunity.

4.3.2.2 Innovation of Meaning

So, not only technology innovation but shape, scenario and narration can suggest us new ideas and products destined to revolutionize the world of production and become best sellers on the international market: from the lunar exploration and Neil Armstrong's lunar boots derive the sneakers, with strong power of absorption thanks to the technology transfer. At the same time looking from another point of

Fig. 4.5 André Courreges has been inspired by Space translating the influence of microgravity through new visions for haute couture. This *polaroid* interpretation is part of an article written by the author. Credits by Room. The Space Journal

view, the Moon landing has had also the ability to inspire great ideas, like the Venetian entrepreneur Giancarlo Zanatta, who in 1970 wanted to clone the footprint of Armstrong's foot to make the mythical snow boots *Moon Boot*, an interesting case study of spin-off related to innovation of meaning, innovation of emotional language, and innovation of usability of the product, and not just to innovation of technology (Fig. 4.6).

"The America is thrilled and celebrates itself for the conquest of Space. Before me—tells Giancarlo Zanatta in an interview—a giant poster of Neil Armstrong walking on the Moon. It's beautiful, it is strong: the man seems to come out from the image. I cannot take my eyes off those boots so special and from that imprint. Well, it happened at all of us if we think about it: the world's attention was not directed precisely at the foot of the astronauts? If we strive to remember, each of us will think of Armstrong descending the ladder, the gait clumsy and bouncy, the famous phrase *a big step for mankind*, witnessed by a so wide and clear footprint on the dust from between the craters. Everything is focused on those feet. Here the bulb lights up. Why not copy those funny boots and make snow boots that leave a similar imprint on the snow? I come back to Giavera del Montello, in my factory,

Fig. 4.6 *Moon Boot* advertising campaign clearly displaying the inspiration from Armstrong's first lunar footprint as well as the innovation of meaning. This *polaroid* interpretation is part of an article written by the author. Credits by Room. The Space Journal

and decided to try it. We are in three on this product and rely on a designer for the logo. The only complication, if I may say so, is the sole that will have to reproduce the footprint on the Moon effect, actually later modified to make it more adherent to the slippery surface of the snow. Not everyone in the company believes in this attempt. Someone says: but where do we go with *this slipper*? Resoundingly contradicted. A few days after the launch at a trade show in Germany, begin to rain orders. Hundreds. Thousands. Forty years later, and that is today, they come to count 23 million pairs sold. Thanks to the *Moon Boot*, Tecnica Group has risen to worldwide fame" (Trabona 2009).

In the new snow boots, the evocation of the great conquest is a very strong appeal to consumers; it is not only a result of an unaware marketing, but also of a strong product innovation because the *Moon Boot* was released using nylon fibre, the maximum in terms of modernity during the Seventies, in a time in which all the mountain shoes continue to use traditional raw materials: the skin, the fur of animal origin. And another big innovation is the introduction of the colour, with a brilliant colour palette, and this is revolutionary. Those are the times of Pop Art and Andy Warhol, who spread a new vision through presenting and valorizing consumer goods as powerful visual communication signs, so important to become privileged subjects of the art system; in this international context, a company which proposes a new shoe coming out from the eternal brown leather is a striking thing! But that's not all: with ambidextrous footwear and without size constraints customers feel more free and light. The *Moon Boot* can be considered also one of the first examples of *democratic design*: it is at the feet of movie stars and queens, but also of all of us, thanks to the simple construction and relatively low costs. This case study is an example of a brilliant idea that makes the enterprise, in the sense of economic activity, an adventure almost poetic, and a product of a museum piece: in 2000, the Louvre has chosen Tecnica's *Moon Boot* as one of the hundred objects of twentieth century which is most representative of the history of world costume. And *Moon Boot* is one of the few trademarks which is mentioned in the Italian dictionary: a so successful product able to become a common name, a word in current use.

The inspiration from Armstrong's footprint (Jenkins 2012) is a transfer of a unique image that reminds us to an extraordinary human challenge, and for that reason, this shape assumes an evocative meaning and a strong power on our emotional choices. The spin-off concerning the emotional language and the usability are harder to identify because they are often based on different factors that need to be integrated together, giving a new interpretation of their meaning.

I experimented this kind of spin-off about usability transfer during the feasibility study carried out for the project *VEST*, described previously in this chapter, retrieving and making use of operating schemes that already exist in familiar clothes on Earth, and transferring them to new garments for Space, which have different needs and requirements hard to imagine considering confined spaces and microgravity conditions, as those connected with body alterations and new kinds of postures and movements on board. I began identifying new wearability parameters to cater for the astronauts' discomforts, creating a collection of garments specifically tailored for the Neutral Body Posture (NBP) that astronauts assumed in Space, which is very similar to a person's posture underwater, with their limbs bent, knees and elbows tending to move upwards and head tipped forward. My surveys carried out during design research have shown that the NBP is, in fact, very similar to the posture adopted by snowboarders, and so I decided to transfer certain functional adjustments designed to make clothing more comfortable for snowboarders—in terms of the cut, stitching and tailoring of garments—to the clothing system for astronauts (Dominoni 2005). In this case, the language of usability allows to transfer conventional outfits that are easier to handle, in another field, in this case, Space, unfamiliar and difficult to manage.

The fashion designers have been inspired by Space envisaging new scenarios explored by the first human space missions and the astronauts' experiences. The microgravity effect, with the possibility to fly and move without to feel weight, and the possibility to look at our planet from another and external point of view, are the two most innovative and extraordinary conditions able to generate visual suggestions that can be translated into shapes, colours and materials: rounded volumes, optical white, and silver nuances as brilliant lunar powder, mirroring metals and synthetic fibres.

4.3.2.3 Fashion in the Age of Technology

The languages and meanings of products do not only derive from research into applications, but they also follow the same kind of workings as the world of technology that cuts across various different sectors to generate spin-offs. Studies into technology management have shown that innovation often comes from a recombination of existing pieces of knowledge. Henri Poincaré talks about creativity as the ability to create useful new combinations of existing elements and claims that the intuitive way of recognizing the usefulness of a new combination is *that it is beautiful*. Of course, he is not talking about beauty in a strictly aesthetic sense, but as something related to elegance as mathematicians understand it: harmony, simplicity of signs, practical correspondence to purpose (Dominoni 2008).

This definition applies to the sciences, arts, technology and also Design. Indeed, it seems to overlap with the design specific capacity to create innovation through transferring of contexts, possible uses, technology and materials into a range of different sectors. This familiar practice among designers is generally referred to by the expression *spin-off* to indicate the design process leading to the creating of useful new combinations by combining existing elements. Hussein Chalayan, Alexander McQueen and Iris Van Herpen are certainly the contemporary fashion designers that have better explored the iconic world of Space looking at technology and smart materials as a plus value to enrich their design proposals and translate them into *science experiments*. British cypriot fashion designer Hussein Chalayan is the master of metamorphosis.

The designer views technology as a *way of pushing the boundaries of what's possible*, and his dedication to sartorial innovation was particularly evident in his fashion show in 2007: he presented a suite of computer-operated *Transformer Dresses* which morphed through over a century of silhouettes reminding to solar panels of satellites. Worn by models stood on the runway, the micro-motor-powered garments creeping across their wearer's bodies as sleeves shortened and hems lifted. In a poetic paradox, the actual aesthetics of these innovative pieces were inspired by Space mixing old-fashioned, with Victorian styles morphing into flapper silhouettes. Hussein Chalayan paid homage to the digital era with his *video dress*, which was made from led lights and played pixel cityscape scenes. Another british fashion designer, Alexander McQueen, has caused a stir when reflecting and prophesying about the future. During his time at the helm of Givenchy in 1999, he presented in the fashion show an art performance with robots spraying colours on the white dress of the model in the centre of the space, who was turning on herself. The techno revolution of Alexander McQueen takes inspiration from Space, science fiction and robots: his models are like cyborgs with their bodysuits covered in neat configurations of multicolour led lights mounted on artificial skin made of moulded transparent Plexiglas® to create body-hugging bodice, which conjured visions of a human circuit board, with each model wearing a 12-V battery pack on her back. Iris Van Herpen, the celebrated *fashion alchemist* who was the first designer to create a 3D printed dress, has long explored the dynamic relationship with science and the digital world. Her visionary brand of sartorialism reached a climax with her human installation *Biopiracy* in 2014. In an uncomfortable voyeuristic display, models were vacuum-packed in plastic, similar to a piece of sous-vide supermarket meat, and suspended several feet above the ground. It wasn't the first time the dutch fashion designer had turned fashion show into a *science experiment*: her haute couture show *Voltage* in 2013, began with lightning like electrical flashes emanating from a model standing on a pedestal.

Always more often the fashion system looks at innovation as a big challenge and a great opportunity to experiment new ways to create garments and new language's codes that could be inspired by the technology transfer. Fashion designers try to interpret this new dimension where all living with managing how to integrate these new notions of digital, virtual and cyber with our real life. If traditionally the distinction between the haute couture and prêt-à-porter was based on the handmade and the machine-made, recently this distinction has become increasingly blurred as both disciplines have embraced the practices and techniques of the other. Today we can choose several handmade couture items, featuring techniques such as embroidery, pleating and lacework, but also machine-made dresses, tailored as handmade, thanks to new technologies like laser cutting, thermo-shaping and circular knitting, or items digitally produced by 3D printing: Chanel couture collection 2015 proposed the prime examples with rich tweed suits woven through 3D printed lattices, and at the Metropolitan Museum of Art in New York the Costume Institute exhibition theme for 2016 was *Manus x Machina: Fashion in an Age of Technology* focused on the dichotomy between handmade haute couture and machine-made fashion.

4.3.3 Project Development of Couture in Orbit ESA-POLIMI

The project *Couture in Orbit ESA-POLIMI* was initially developed inside my Course at the Master of Science in Design for the Fashion System, at the School of Design, Politecnico di Milano, where I worked six months with the students to arrive at the definition of many concepts of garments inspired by space scenario and enriched with space technology and materials given by the European Space Agency (ESA) and its technical partners.

During this first phase, we were constantly in contact with the leader of *Couture in Orbit* in ESA, Rosita Suenson, who came to Milan for the kick-off of the project to give us information and contact to develop the project. We have been the go-between, with the students on one side and the industries involved by ESA on another, among which Rina Consulting, the Italian partner of ESA for Technology Transfer Solutions, and Extreme Materials, a small company specialized in textile-related technologies, which has in the past included such innovations smart materials embedded in textiles and composites. At the end of this firsy phase we presented the concepts at the Science & Technology National Museum *Leonardo*

Da Vinci in Milan to ESA experts, and we selected together with Rosita Suenson the three best ideas to implement for the capsule collection. For each project we had to design and complete two outfits, a total of six outfits, to present at the fashion show at the London Science Museum on 25 May 2016 (Fig. 4.7).

The second phase of the work continued as a research project with the support of our Fashion Lab, for the realization of the whole garments, and finished with the photo shooting at the Immagine Lab, both to support the activities of our School of Design at Politecnico di Milano.

ESA provided space-certifies textiles for the students to use in the project, which was supported by the companies 37.5, Bionic Yarn, and Sympatex. The company 37.5 produces fabric technology invented in the 2000s by Cocona, Inc. to make use of the waste of coconut husks from the food-service industry, which would otherwise go to a landfill. The manufacturer burns the coconut, and the activated charcoal fibres are embedded in yarn to increase the material's surface area, thus creating a moisture-wicking effect. Bionic Yarn is focused on recycling resources we consume on Earth in order to have a positive impact on society and our natural environment. Turning things we discard into raw materials we can be proud to use in our everyday lives. They make high-performance yarns mixing naturals and synthetics fibres made with recovered plastic (RPET). As one of the world's leading suppliers, Sympatex® Technologies has been a byword for high-tech functional systems in clothing, footwear accessories and work-wear for decades. In addition to its standard functions (optimal breathability, 100% water-proofness, 100% wind-proofness), the Sympatex® membrane is also 100% recyclable, PFTE-free and PFC-free as well as Oeko-Tex® Standard 100 certified and Bluesign® approved. At Sympatex®, superior quality and well-thought-out technology are essential building blocks for optimal performance and comfort. With this in mind, Sympatex® developed a non-porous membrane consisting of billions of hydrophilic molecules which uses a physiochemical process to rapidly transport moisture from the inside to the outside. It ensures that the body always remains dry—even in extreme situations. Not only that the membrane also provides effective protection against rain. Unlike standard micro-porous structures, the Sympatex® membrane retains maximum functionality even after activity in continuous rain or washing as there are no pores that can become blocked.

In addition, we received from ESA other materials made of Multi-Layer Insulation Film (MLI), space blanket of heat-reflective foil in aluminium, very light with low-weight, used for many applications on the spacecraft, such as structural members, instrumentation, antennas, radiators, wiring harnesses and sunshield covers.

Studying product design from the viewpoint of fibres and textile's structures draws attention to the fact that material is one of the key factors in any project, capable of stimulating structural, formal and aesthetic experimentation, and also of generating spin-offs and applications in a range of very different sectors. Characteristics of lightness and flexibility derive directly from research projects and applications in the aerospace industry, where the cost of transporting material into orbit (and hence weight) is incredibly high for the amount of energy used.

Fig. 4.7 The final catwalk of *Couture in Orbit* collection at the London Science Museum, which integrates space technology into garments to be worn on Earth. Credits by ESA

4.3.4 Learning Goals

The Space imaginary offers fantastic experimental scenarios that can be transformed and reinterpreted by the language of fashion, so apparently distant from the scientific world that revolves around Space, for example, by finding new applications of clothing technology that can improve the comfort of the people who wear them. The fashion system plays an important role in this spin-off process, because it is a catalyst of trends, lifestyles and behaviours, and at the same time is open to the use of new materials, intelligent fabrics and wearables. In this context, *Couture in Orbit* was also an opportunity to reflect on the role of technology in relation to fashion that increasingly considers innovation as a value to be integrated into their production processes, especially if tailoring matrix. Wondering how technological innovation can influence people's behaviours and performances, and how we could become more aware of this process of transformation is the task of Design, and it is characteristic of our School of Design of Politecnico di Milano to create for our students important opportunities for research and comparison with the real world, such as this prestigious project with the European Space Agency (ESA).

Fig. 4.8 *Couture in Orbit ESA-POLIMI.* Could a wearable *Cooling Technology* be modified to gradually deliver fragrances through steam instead of water? Credits by Politecnico di Milano

Our principal aim was to find relations between life in Space and life on Earth connecting the two environments and considering not only technological spin-offs or spin-ins, but also emotional language and usability. We transformed inspirations into shapes and textures by looking at astronaut postures and movements in microgravity and considering their activities in Space in relation to the objects in order to generate new gestures and ideas for our projects. From the aesthetic point of view, we were inspired by space images, choosing colours like silver and white in all the lunar nuances, looking at materials such as aluminium and reflective surfaces, selecting curved shapes and imagining textiles printed with wonderful pictures taken by astronauts looking out of the International Space Station (ISS) towards Earth as well as by satellites, with marvellous colours, patterns and phenomena of light (Dominoni and Quaquaro 2017a, b). But we pushed even further, trying to imagine new uses of space technologies that they do not yet have applications other than the original, but they could in the short period of time become very interesting. Translating high fashion concepts into the language of technology was an essential skill for an endeavour like the project *Couture in Orbit*.

4.3.5 Design Methodology

We started to develop the project considering the aesthetic fascination of Space, the technological and functional aspects of the garments, the emerging scenarios in our society and the new needs, the possible spin-offs between Space and Earth and the most suitable expressive languages to be applied to fashion. We were inspired by the confined environment and microgravity that astronauts live on board the International Space Station (ISS), analyzing their activities in orbit and the relationships with objects, which without weight generate new behaviours and gestures following the *Use and Gesture Design* (UGD) methodology.

The materials that attracted us the most are reflective and very light, made of pure aluminium plywood alternated with white nets similar to gauze (but much more resistant) that are used to make satellites. The lunar colours of white and silver alternate with colourful prints inspired by the images of the Earth, thanks to the satellite network that, depending on the size scale, change patterns and assume both geometric and organic configurations.

We chose technologies developed by the technology transfer system ESA Technology Transfer Program (TTP), such as the *Cooling Technology* used to keep the temperature constant in astronaut suits, and subsequently chosen as a spin-off from McLaren company for protective mechanics suits. But we went even further, trying to imagine new uses of space technologies that do not yet have applications other than the original, such as some *ESA Patents of antennas*, but that could in the short term become very interesting if printed with conductive ink directly on fabric. The attention to technological aspects did not divert us from the goal of also focusing on aesthetic qualities: I am convinced in fact that *innovation* and *beauty* are the most important keywords to create a collection of clothes able to integrate space technologies with a *glamour* vision of fashion.

We consider multidisciplinary a great value in which each student collaborated in a unique teamwork with different backgrounds, knowledge and talents to achieve the same goal. The greatest opportunity was to research and work for such a prestigious client as ESA and all the students were thrilled for this experience. The students have sought a clear explanation of how these technologies are used in practice, and in other cases consulted with the companies involved into the project to find original solutions and developments to the problems they have posed. We wanted to encourage the students to come up with apparently crazy applications—to stay as crazy as possible—while we were the ones with our feet on the ground, to take into consideration the feasibility of implementation of ideas in a concrete, workable way. For instance, could a wearable *Cooling Technology* be modified (Fig. 4.8) to gradually deliver fragrances instead of water? And could an innovative *ESA Patent for a 3D antenna* can be printed directly on textiles with conductive ink?

4.3.6 *Couture in Orbit Projects*

The three winning projects *Therapic Garments 23.44°*, *Tourist in Space* and *Food Keeper* were selected by a jury composed by Rosita Suenson by ESA together with other experts and project managers of the European Space Agency, the responsible of the industries Rina Consulting, the Italian partner of ESA for Technology Transfer Solutions, and Extreme Materials, Benedetto Quaquaro and I. Our team followed the students from the first ideas to the realization of prototypes and declared themselves very satisfied with the results. The research group that created the capsule collection was international, made up of students from all over the world—England, Brazil, France, Argentina, Serbia and Colombia, as well as Italy—who share and exchange cultures, different ways of living and thinking, enriching the quality of the projects.

4.3.6.1 Therapic Garments 23.44°

The number *23.44°* corresponds to the inclination of the axis of the Earth that determines in the inhabitants of the northern regions a biochemical disorder called Seasonal Affect Disorder (SAD) caused by the lack of sun—and able to generate anxiety and depression—and which can be compared to the alteration of circadian rhythms of astronauts confined aboard the International Space Station (ISS) without natural light (Moore et al. 1996). The concept of this project focuses on body alterations and the study of therapeutic essences.[5] The principal aims are to respond to these needs by acting on the possibility of stimulating the mind and body and restoring the biochemical balance with fragrances inserted in tubular structures integrated into the clothes that spread in the air through steam.

The idea is inspired by space *Cooling Technology*, developed by ESA Technology Transfer Programme (TTP) and originally used in the astronaut's Extra-Vehicular Activity (EVA) under suits to keep the temperature constant thanks to heated water that flows in a system of tubes sewn on the inner fabric of the garment. The most famous case study is McLaren mechanics' suits, made by the Italian fashion manufacturer Karada together with the fashion designer Hugo Boss. The challenge was to produce a thermoregulating garment, offering fire protection and a comfortable working temperature for the whole team servicing the car, whose overalls have the same safety standards as the pilot's. The ESA TTP offered the solution: 50 m of plastic tubing, 2 mm wide, developed for astronaut's suit by the Canadian company Med-Eng, and installed by Karada in 55 overalls. The result is a miniaturised air conditioning system, offering maximum comfort when working under extreme heat.

[5]The project *23.44° Therapy Garment* was developed by R. Fustinoni, G. Presti and S. Ward under the supervision of professors Annalisa Dominoni and Benedetto Quaquaro of Politecnico di Milano, and with the technical support of Andrea Ferrari by Rina Consulting, the Italian partner of ESA for Technology Transfer Solutions, and Ettore Rossini by Extreme Materials.

Fig. 4.9 *Couture in Orbit ESA-POLIMI*. The tubes of the *Therapic Garments 23.44°* flow on the dress creating an arabesque that paradoxically could have a strong identity even without the technology in use. Credits by Politecnico di Milano

In the 23.44° project it is foreseen a plastic tubing structure, surrounded by a tubular knitted structure and integrated into garments, which releases substances through steam, instead of water like in the traditional Space *Cooling Technology*, and distribute themselves on the surface of the garments. The tubular circuits—made of flexible plastic and wrapped in a tubular 3D coloured mesh of Extreme Materials company—create strong and recognizable organic signs on clothes and provide micro-holes that allow fragrances to envelop the body. The substances are inside small tanks in a solid form. Together with pharmaceutical substances, it is possible to join perfume essences as orange and eucalyptus to increase dopamine levels making the wearer feel more alert and energetic, but also essences of vanilla, anise and lavender that help the release of hormones such as oxytocin, serotonin, and norepinephrine, for relaxing effects.

Sculptural and organic volumes define dresses that invite those who wear them to experience an *intra-body transformation* in which smell plays the main role to stimulate mind and body, reactivate memories and stimulate emotions. The choice of white colour is inspired by the essential white of the Extra-Vehicular Activity (EVA) suits that astronauts wear in Space, while the tubes are coloured and in accordance with the performances induced by the essences.

This project introduces also another important aspect using Design approach in Space: the value of the beauty beyond the technology described in 1st chapter *The Strategic Role of Design for Space*. In addition to the choice to make the technology visible, by putting the tubes out of the dress and not inside, as in the space Extra-Vehicular Activity (EVA) under suit, we made sure that the tubes flowed on

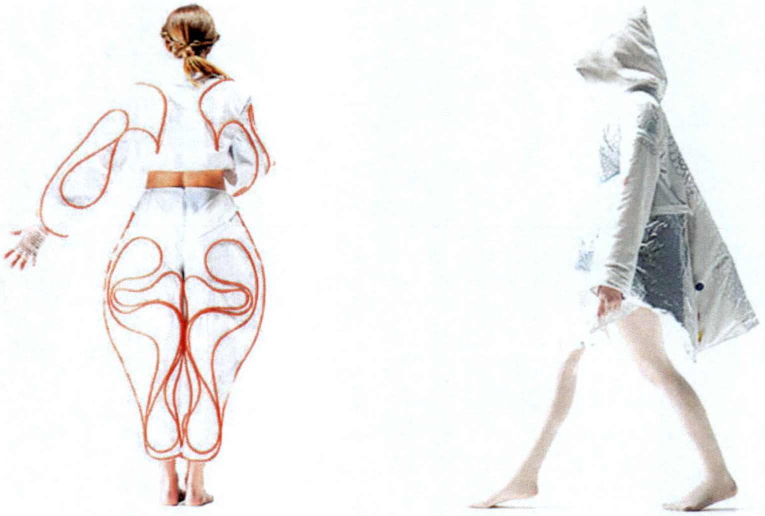

Fig. 4.10 *Couture in Orbit ESA-POLIMI*. Design is a bridge between technology and beauty: *Cooling Technology* and aluminium Multy Layer Insulation Film (MLI) inspired *Therapic Garment 23.44°* (left) and *Tourist in Space* (right). Credits by Politecnico di Milano

the dress of the *Therapic Garments 23.44°* creating a drawing, an *arabesque*, that would have had a strong visual impact, and that paradoxically could have had an identity even without the technology in use (Fig. 4.9).

This is a case study that shows how through Design we can transform technology inputs in beauty driving both the languages of science and style (Fig. 4.10).

4.3.6.2 Food Keeper

This capsule collection is inspired by the food in Space and the types of packaging specific to the confined environment and microgravity that contain foods often in the form freeze-dried and unattractive.[6] Astronauts are forced to eat in almost perpetual motion, floating on board the International Space Station (ISS) without having the possibility to remain motionless, if not anchoring to the structures provided on board the habitable modules. *Food Keeper* offers clothes that allow you to carry and store your food carrying your meals wherever you go and to find the pleasure of eating it, even if on the move, when we walk and move in the city.

[6]The project *Food Keeper* was developed by A. Laurentin and C. Martinez under the supervision of professors Annalisa Dominoni and Benedetto Quaquaro of Politecnico di Milano, and with the technical support of Andrea Ferrari by Rina Consulting, the Italian partner of ESA for Technology Transfer Solutions, and Ettore Rossini by Extreme Materials.

Fig. 4.11 *Couture in Orbit ESA-POLIMI*. The lunar craters of the *Food Keeper* capsule collection are made by 3D knitted fabric that allow to create garments with strong and defined volumes. Credits by Politecnico di Milano

Each dress has a different function determined by various types of compartment pockets in very large compartments: some are designed to store food and others to carry water in the amount necessary to not have to skip meals. The space style of the clothes recalls the 1960s Space Age by Pierre Cardin and André Courreges, the shapes and volumes of the pockets are reminiscent of lunar craters and are made of insulating material padded and lightweight—Multy Layer Insulation Film (MLI) aluminium used by European Space Agency (ESA) for various applications in spacecraft, instrumentation, wiring, structural elements and reflective parts of satellites—to be able to contain, transport and store food in the best possible state. The use of 3D knitted fabrics by Extreme Materials has allowed to create garments with strong and defined volumes (Fig. 4.11) that alternate with transparent and light crinolines on layers of MLI making glimpse the reflections of silver and gold light of the material.

Another silver tubular mesh covers the pipes that carry water to drink and maintain the constant temperature—inspired by the ESA *Cooling Technology*—creating at the same time aesthetic and structural decorations on the clothes. In this case, the space *Cooling Technology* is interpreted in another opportunity: the idea is to drink water through the tubing structure that decorates the outfits with a double intent: aesthetic and functional. A small tank can be filled and carried in a very light way. MLI aluminium film substrates are perfect to conserve food qualities in the pockets while moving across the city.

Fig. 4.12 *Couture in Orbit ESA-POLIMI*. The shooting backstage of *Tourist in Space* capsule collection at the Immagine Lab of Politecnico di Milano. It is possible to see the pattern of the *ESA antenna* printed on textile with conductive ink. Credits by the author

4.3.6.3 Tourist in Space

The project looks at the need to carry payloads and equipment in orbit on the International Space Station (ISS) that have the least weight and volume possible and can be used and stored quickly and easily by astronauts, space travellers, tourists, living in an extreme and difficult to manage environment. As well as astronauts, travellers face similar situations, although not so extraordinary: weight and luggage volume restrictions, climate changes, etc. Thinking about the comfort and the transformability necessary to the clothes we wear when we travel and want to keep a light luggage has generated a system of garments that bend – following the japanese technique of origami just like satellites and solar panels—and also change shape and function through a zip system that allows you to switch from outerwear to a dress by removing and moving the parts that compose it.

The foldable clothing system suits the tourist and the traveller who has to change the look during the day passing from a flight by plane to a business meeting and maybe even a dinner party keeping the same dress, even if with different configurations, included a sleeping bag stuffed with lightweight material, as aluminium Multy Layer Insulation Film (MLI), and thermal insulation used by the European Space Agency (ESA) for different applications in spacecraft, instrumentation, wiring, structural elements and reflective parts of satellites. Sympatex®, a high-performance technical fabric on which are printed conductive ink antennas (Fig.4.12)—inspired by an ESA patent—has been chosen to create transformable

Fig. 4.13 *Cooling Technology*: the picture shows how the cooling tubes are embedded into a 3D textile. In *Couture in Orbit ESA-POLIMI,* we imagine to make the steam also flow beyond the water as in the Extra-Vehicular Activity (EVA) under suit. Credits by Extreme Materials

clothes, which, in addition to forming a strong and recognizable graphic pattern, allow localization, the transmission and reception of signals.

The white background contrasts with the colourful printed parts with images of the Earth seen from Space thanks to satellites, while the colours of the hinges suggest how to use the garments and transform them. The inspiration from satellites brings another value to this project: the possibility to have a localization system printed on textile with conductive ink which works as an antenna. The ESA Patent of a 3D antenna, *Optimized Antenna Elements Position and Dimensions*, offers a method and software product with which antennas are designed.[7] The configuration of the antenna is made of little 3D cones that looking from the flat side creates a pattern of concentric little circles similar to a flower.

In this case study, like in the project *Therapic Garments 23.44°* ispired by the space Cooling Technology (Fig. 4.13), the idea to translate these cones into the same configuration of the antenna, but printed in 2D with conductive ink, is a good example to demonstrate the power of beauty, transforming technological pieces—integrated together in that configuration just following functional principles—into a significant drawing (Fig. 4.14) that gives a strong identity and recognition to the garment, as explained in the paragraph *A Bridge Between Human Science, Technology & Beauty* in first chapter *The Strategic Role of Design for Space*.

[7]ESA Patent PAT619, *Optimized Antenna Elements Position and Dimensions.*

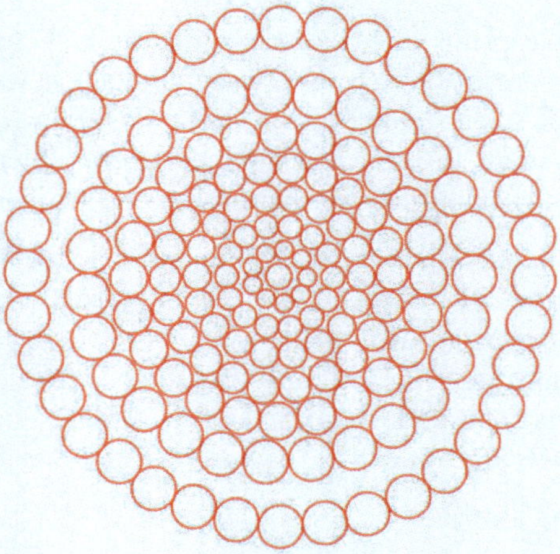

Fig. 4.14 *ESA Patent of Optimized Antenna Elements Position and Dimensions*: the picture shows a circular layout of array designed accordingly to the present methodology and employing radiating elements with different apertures. Credits by ESA

4.3.7 Learning Outcomes

Couture in Orbit ESA-POLIMI allowed students to discover the potentialities of Space as a source of inspiration to reach new ideas and to imagine new scenarios on Earth in which Space Technology can be applied in different ways to increase the quality of life, combining aesthetic values with functional aspects and society's needs. They focused on a *design vision* with an *out-of-the-box approach*—like an astronaut looking at Earth from Space—but developed a real capsule collection starting from designing concepts to the realization of the prototypes.

An important part of the project was the relations of students with ESA's partners and companies of the private sector during the development of the design concepts that allowed them to be more conscious of space technologies and smart materials applied in the capsule collection. In addition to the project, they learnt how to create and organize a catwalk and an exhibition participating to all the phases of the event *Couture in Orbit* at the London Science Museum and going in depth into this experience with a strong enthusiasm.

4.3.8 Results Dissemination and Exploitation

The results of our capsule collection *Couture in Orbit* from the Italian side *ESA-POLIMI* were spread through the most important newspaper and magazines crossing the fields of science and fashion, from *Vogue* to *Il Sole 24 Ore*, and we received many invitations to present our project during international events, exhibitions, symposia and scientific TV programs among which.

Space for Inspiration, event-talk by ESA, London Science Museum, 2016;
Couture in Orbit ESA-POLIMI, event-catwalk organized with ESA-ESRIN for the *European Researchers Night*, Rome 2016;
Fashion in Space, lecture at the National Conference *Space4Sustainability*, organized by the Associazione delle Imprese Per le Attività Spaziali (AIPAS) Macro Museum, Rome, 2016 are some.

We received also the awards:

- *ADI Design Index 2017*, the prestigious selection of the best products of the year by *Compasso D'Oro* [7];
- *Lombardia's Excellences Award 2017.*

4.4 Space4InspirAction. A Space Design Course with ESA

Space4InspirAction (S4I)—created in 2017 by Annalisa Dominoni and Benedetto Quaquaro and now at its 4th edition 2020—is a Space Design Course, Master of Science in Integrated Product Design, at the School of Design, Politecnico di Milano. It is the first and unique international Course of Space Design in the world recognized and supported by the European Space Agency (ESA) through experts and scientists who suggest and then deepen the project themes in line with the objectives of the space Agencies strategic programmes. The collaboration between the S4I Course and ESA is not a spot project, but the beginning of an ongoing agreement about an *academy—research—design* process to ensure the development of innovative ideas in which the contribution of Design discipline is fundamental and strategic to generate disruptive visions of habitability and equipment solutions to increase the wellness in Space, with particular attention to the *new human factors*[8] and interactions between operator, tools and environment (Dominoni et al. 2018). We have every year international students coming from all the world who contribute to create a multidisciplinary working group thanks to different background, cultures, curricula and know-how. There is a big selection and only the best students can participate, and in the maximum number allowed, not more than sixty.

[8]The *new human factors* are treated in third chapter *Research and Design for Space Life*.

When we created the Course with the support of ESA, we decide to take inspiration by the name of the international congress, Space4Inspiration, to which we have been invited, as guest speakers by ESA, to lead a lecture of Space Design in 2017 at the London Science Museum. Our idea to add *Action* to the word *Inspiration* wanted to underline the design soul *in action* of the discipline in Space Design. We wanted to bring out how we can both design for Space and for astronauts, but also how Space can be a great inspiration to generate terrestrial innovations through space technology transfers, but also behavioural and visual spin-offs and spin-ins.

Space Design is a new discipline that is being formed due to the strong attraction that Space has always generated on society, thanks to its narrative power, and which will have a decisive influence on people's behaviours and performances in the near future. Design can speak different languages, combines *science and beauty*, becomes a mediator between technology and needs through projects that generate new scenarios and new paradigms with which to confront.

Space can be a great inspiration to create *design-driven* projects that can inspire companies which look with interest in innovation and research to find new products.

Moreover, the International Space Station (ISS) can be seen as the most *extreme example of sustainability* that the human being has managed to achieve because it can be considered a *living organism* capable of recycling 100% of its resources. Besides, it is the only international experiment, visible to everyone, that has managed to unite different cultures and races in a project of collaboration and scientific progress for the whole world which goes beyond the divisions between countries.

The Course *S4I* is a unique opportunity to develop a strong capacity of *visioning* looking at Space as an inspiration to innovate and design new projects of Space habitats and tools, but also new scenarios that could help people to live better in a sustainable way. Furthermore, Space allows students to develop creativity by confronting with confined and microgravity environment, that is not part of our common experience. I believe Space innovations will have a strong influence on how people behave and perform, and the role of Design is to *humanize technology* to meet the needs of a sustainable society. I also believe Space can really inspire students and help to understand better how they can be more conscious of the transformation technology brings.

The aim of our Course *S4I* is to create new professional figures which will be able to

- connect both technology and beauty languages,
- increase creativity and visioning,
- find design solutions crossing know-how and research,
- imagine new cultural and business models to increase well-being according to sustainability.

The partnership with ESA foresees the participation of experts and scientists who define together with us the themes of the projects to develop together with students, according to the strategic programmes of the space agencies, and integrates their experience and know-how with lectures useful to set up the base of the research and the development of the projects. ESA's support is a big value because a theme like Space allows students to develop creativity with intense visioning activities by confronting with confined and microgravity environment that is not part of our common experience. Thanks to ESA we are able to compare our design vision with real conditions and requirements and find innovative solutions designed for Space, but that can be also transferred to daily life on Earth becoming spin-offs.

The Course *S4I* has a dual purpose: on the one hand pushes students to get inspired by Space and define new visions; however, starting with real conditions and requirements, on the other to imagine new spin-offs, or terrestrial applications, that can be transferred to our daily lives to improve them. For this reason, design companies from the private sector are involved in the projects, together with space companies, with the aim to translate space inspiration and vision into spin-offs generating new concepts of products. It is our aim to enhance the high quality of the projects developed during *S4I* with the exploitation of the results through media press and exhibitions.

The frame of *Space4InspirAction* Course is composed of two parts which interact with each other:

- a theoretical and explorative part which provides lectures with ESA experts and scientists as well as engineers by space industries who integrate their experience and know-how useful to the development of the projects;
- a practical part, the real project, developed with professors and supported by ESA and space industries in which is required a disruptive capacity of visioning.

The sequence of activities is planned to introduce progressive levels of complexity and to give students the opportunity to use previous knowledge and that they are learning in parallel in the theoretical integrated part with the aim to increase their capacity to manage information with different skills.

The topics of *Space4InspirAction* Course changes every year, following the ESA's space strategic programmes, with the aim to find connections between *living in Space* and *living on Earth* and suitable solutions for two both, looking at astronauts activities in microgravity and the alterations of body postures and movements in relation to their objects that can generate new gestures and ideas for our projects. Space knowledge, space research results, experimentations, new materials and technologies suggested by ESA are important drivers to increase the innovation value of the projects.

4.4.1 Design Methodology for Space

The unusual and extraordinary environment offers the students the possibility to *think out-of-the-box*, like astronauts' experience looking at the Earth from another point of view, and the chance to increase the creativity and the ability to imagine and design new objects and tools starting from a perspective completely unknown. *Microgravity* and *confinement* are the principal keywords of Space to inspire new projects, stimulate reflections and connections between Space and Earth and find suitable solutions for both environments.

Through a new methodology[9] that I have developed and called *Usage and Gesture Design (UGD)*, and that is specific for Space Design, emerges as the figure of designers is not limited to creating objects, environments, architectures, but can be extended to the suggestion of new *behaviours* and *gestures* of astronauts in relation to the use of designed objects, environments and architectures.

Multidisciplinary approach is fundamental for the creation of new professionals who are able to speak different languages of the project and that is why we use to involve in the Course *S4I*—a part ESA experts of different areas—aerospace, mechanic and mechatronic engineering students, who work in groups with our design students, including product design and interaction design, to ensure the feasibility of new creative and disruptive ideas, and increase the skills of all students.

Of course, all the project we developed would not be possible without the support and *participation of ESA and companies*, both in the space sector and private, to face real contest and real problems and find disruptive ideas but based on feasible know-how and technology. This is an important part of the project that push students to be confronted with ESA's partners and companies during the development of the projects, because they can become more conscious of space programmes, technologies and challenges crossing different knowledge. We think innovation can arise by exploring many fields, better if they are very distant, and produce disruptive ideas as the case studies reported in first chapter of this book *The Strategic Role of Design for Space*.

A last point, that is not properly a methodology, but a practice that we pursue with conviction is the importance of dissemination and exploitation of the results through scientific publications, newspapers and magazines with big circulation, radio and tv programmes as well as exhibitions.

Beyond these methodologies, there are others integrated in the path of Space Design which belong to the Design discipline and that are very useful in Space. Here I summarize briefly all the approaches, practices and suggestions to face the development of the projects, starting from Design and ending with Space Design's methodologies:

[9]The methodology *Usage and Gesture Design (UGD)* is treated extensively in this book especially inside first and third chapter because is the base of the Space Design discipline.

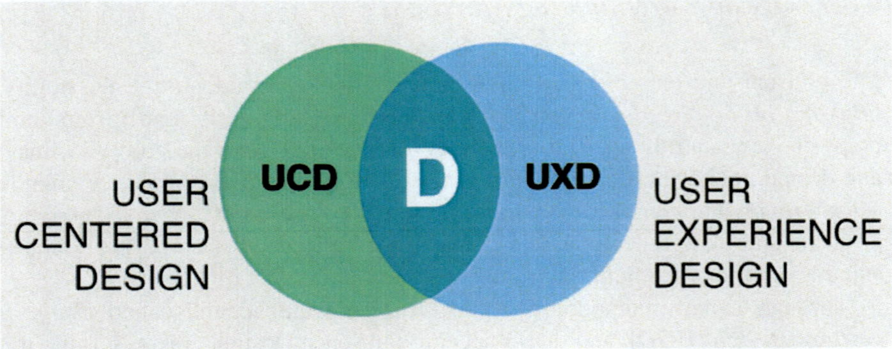

Fig. 4.15 *User Center Design (UCD) + User Experience Design (UXD)*: the most significant characteristic that distinguishes the designer's approach is a prominent attention to the *users* of the products, their *needs* and *experience* expectations. Credits by the author

- the relevance of *thinking out-of-the-box* for visionary projects: the unusual and extraordinary environment offers students the possibility to think out-of-the-box, like astronauts experience looking at the Earth from another point of view, and the chance to increase the creativity together with the ability to imagine and design new space environments and tools, starting from a perspective completely unknown (Dominoni 2015);
- Design is a *problem creating* discipline: I like to define the designer a *problem creator*, the opposite of a *problem solver*, which generates innovation and increases people's well-being, and this is true both in Space and on Earth. Design is useful if it finds new problems to solve, if it creates something new that nobody has ever thought before. If an astronaut lives and works in a comfortable environment, with equipment and facilities designed properly to be used in Space, its performances increase as well as the quality of the activities, individual and collective, and can be determinant for the success of a mission;
- the *new human factors*: the boundaries of intervention areas for Design seem to converge into a new methodology with a background *halfway between design and human sciences* characterized by an increasing attention towards the user, and the value relative to the use and to the emotions aroused by the product on one hand, by a strong involvement in the processes of technical scientific innovation on the other (Green and Jordan 2019);
- *User Center Design (UCD) + User Experience Design (UXD)*: the most significant characteristic that distinguishes the designer's approach is a prominent attention to the *users* of the products. The designer must interpret, by means of the formal and symbolic qualities of its projects, people needs, expectations and desires, and at the same time, elevating the degree of their general *well-being*. Design places the human beings at the centre, with a *User Center Design (UCD)* approach, trying to interpret their needs in the most sustainable way. Going beyond, Design creates new *experiences* for people, with a *User Experience*

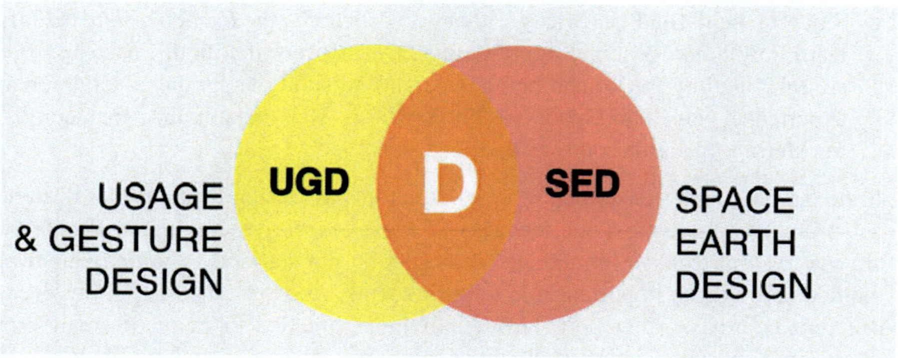

Fig. 4.16 *Use and Gesture Design (UGD) + Space & Earth Design (SED)*: applying a *UGD* approach means projecting the features of the future object on the scene of its possible uses and visualize them in action. Credits by the author

Design (UXD) methodology, imagining how it could make more pleasurable and unforgettable the human life (Fig. 4.15);

- focus on the design process: during the design process we have to consider, as usual, all the methodologies which belong to the Design discipline (Dominoni 2010) as *design thinking*, in the preliminary phase of the concept design, followed by *design to cost* and *design to recycle*, during the development of the process to make the project feasible, attractive, affordable and usable;
- *Use and Gesture Design (UGD)*: the most important and most interesting difference between Design for Earth and Design for Space can be identified in the design process, which regards the capacity to forecast the design of use and gesture, visualizing possible human gestures and movements in Space, in confined environments and in microgravity conditions, which cannot be compared to human experience on Earth (Fig. 4.16). Through a new methodology that I have developed and called *Usage Gesture Design (UGD)* we do not limit ourselves to creating objects, environments, architectures, but also, we design new behaviours and gestures of astronauts in relation to the use of the new objects, environments and architectures. Applying a *UGD* approach means projecting the features of the future object on the scene of its possible uses and visualize them in action: in your hands, put on you, in the environment to zero gravity where the object will be used, and how it will be used, in which ways, etc. That means the design activity to create a new object, or an environment, must be contemporary to the forecast of the use of the new object, or environment. The designer becomes a filmmaker, able to create a potential script of the new object, together with the environment, the movements, and the gestures of the astronauts;
- crossover between *Space & Earth Design (SED)*: Design of Space is also a process from Space to Earth and vice versa to generate spin-offs and spin-ins, technology and behaviour transfers, new applications and new products. We push students into a design process in which they experiment a practice crossing;

- different fields and knowledge, that we called *Space Earth Design (SED)* to learn how it can generate opportunities. Considering that both Space and Earth are inhabited by the human being, we can find solutions by the observation and experience gained in Space (technologies as well behaviours) that could be transferred on Earth contexts and vice versa.

If the methodologies described represent the tools to Design for Space, as designers and architects, there are some methods of *active learning* that in the academic field are very useful to have immediate feedbacks by the students. The *active learning* methods provide an important role to the students, who are called to get involved in different contexts. They range from individual practice to pairs or small group discussion on specific topics, to the exploration of topics and problem-solving, making the entire learning experience a collective research, connection and knowledge processing experience. The *active learning* methodologies are also clearly related to the *flipped classroom* approach, which is based on the idea that students explore some content before the lesson which is then given maximum value for the application of knowledge in a dynamic interaction with teachers and peers. Usually, we programme lectures, with ESA scientists and companies' experts, at the beginning of the Course *S4I* to allow students to go in depth into information and tools to approach the projects. The lectures are scheduled to be maximun 45 min each followed by a 30 min session interspersed with moments for questions and for the consoling of the notions just learned through the use of *Socrative*. The professor prepares 3/5 questions on the content explained during the lecture with the aim of consolidating the knowledge and skills constituting the lecture's fundamental concepts to be taken away. Students answer individually or in pairs. *Socrative* allows to visualize the answers to multiple choices maintaining the anonymity of the students and giving us an immediate feedback on the correct transposition of the information, and therefore of the clarity of the contents expressed. This tool is one of our favourite methods of active learning because we found it is a very useful system to have feedback and consolidate knowledge: usually, the majority of students responds correctly and besides, who has the wrong answer, more easily remember the new information. Another method that we use to check the mood of the classroom and know what the students have in mind is *AnswerGarden*, a very simple and minimalistic tool that *photograph the thoughts* of a precise moment of all the participants—that could be after a lecture, an exercise or an explanation of the project organization—takes shape and grows up according to the evolution of the day, the topics of the lectures, the feeling and the emotion of everyone involved. Thanks to *AnswerGarden* the class becomes a *living organism* that reacts to the inputs given by professors, guest speakers and students, and which creates a feedback in movement and open to a continuous transformation during the day: this is an *active learning* experience able to increase the empathy between us and our students that can be associated to the idea of a self-sustaining spaceship, which changes behaviours according to the various environmental stimuli and grows, thanks to the participation of all the actors involved.

Both *Socrative* and *AnswerGarden* are virtual tools that can be used by students in classroom, but that we discovered was very useful also during our last learning experience leading the Course *Space4InspirAction (S4I)* 4th edition in distance due to the Covid-19 pandemic in 2020; we developed projects with a network of students, who are living in different parts of the world, and in which, despite the distance, all the actors collaborated to enrich the experience.

To conclude this paragraph of the *Design Methodology for Space* I would report the data collected by us after the end of the Course S4I 1st edition concerning questionnaires in which we asked our design students to give us feedback on the experience of the new Space Design context; the involvement of ESA and companies both from Space and the private sectors in the development of the projects; the working groups with multidisciplinary teams made of design and engineering students; the impact of our methodologies; the three keywords that in their opinion the peculiarities and the best qualities of the Course *Space4InspirAction* and their suggestions. We raise an interesting list of comments in which they underlined the advantages of this new experience:

1. a Course of Space Design is an added value of the School of Design;
2. ESA and industries support the projects' optimization with integration of technological know-how, and it is a great opportunity to increase knowledge and compare ideas and concepts with feasibility;
3. the multidisciplinary teams organize ideas and concepts in a coherent logic;
4. the collaboration with different professional figures will increasingly characterize future works and this experience allows to open a constructive dialogue of exchange;
5. opportunity to know each other and debunk myths as *the designer is an artist and deals only with shaping things, the engineer is too rigid mentally*;
6. collaboration increases inspiration and multidisciplinary communication;
7. methodologies balance visioning and practice for more inspiring projects.

The most recurring tree keywords to describe the Course *S4I* that the students suggested in the questionnaires were freedom, creativity, experimentation and I think they reflect very well the intent and the spirit that prompted us to create a new discipline in Space Design and to experience during these first four years many issues related to how we live today on the International Space Station (ISS), but also to those that we will soon live on the next space stations and lunar and martian bases, which will open a new era of planetary colonization.

4.4.2 Space4InspirAction 1st Edition 2017

The themes we have chosen during the 1st edition were focused on three topics and involved companies with the specific know-how to help us developing the projects together with the European Space Agency (ESA):

Fig. 4.17 A picture taken at the end of the event organized for the 1st edition of *Space4InspirAction* at *Planetarium Ulrico Hoepli* in Milan. In the first row from the left, the author, Raffaella Pappalardo by ESA and Benedetto Quaquaro. Credits by M. Bergamini

- *Space Food,* considering that nutrition is one of the topics that attract the attention of the space scientific community with the aim to create an autonomous life cycle integrating plants growth in the space habitats; for this project, we chose Argotec, the Italian company which realized and launched a coffee machine on the International Space Station (ISS) together with Lavazza, and which created the *bonus food* for the astronaut Samantha Cristoforetti during the *Futura Mission.*
- *EVA Suit*, that means a suit for Extra-Vehicular Activity (EVA), a very complex project that we could consider like a small space station complete with all the parts and systems useful to maintain the living conditions inside, protecting the human being to the vacuum of Space outside; we have chosen the Italian company Dainese to support us, leader for motorcycle garments and accessories including protective helmets, gloves and boots. In addition to dealing with protective sports clothing, they have a strong know-how in designing space suits for the ISS Intra-Vehicular Activity (IVA) and the Moon working EVA for NASA.

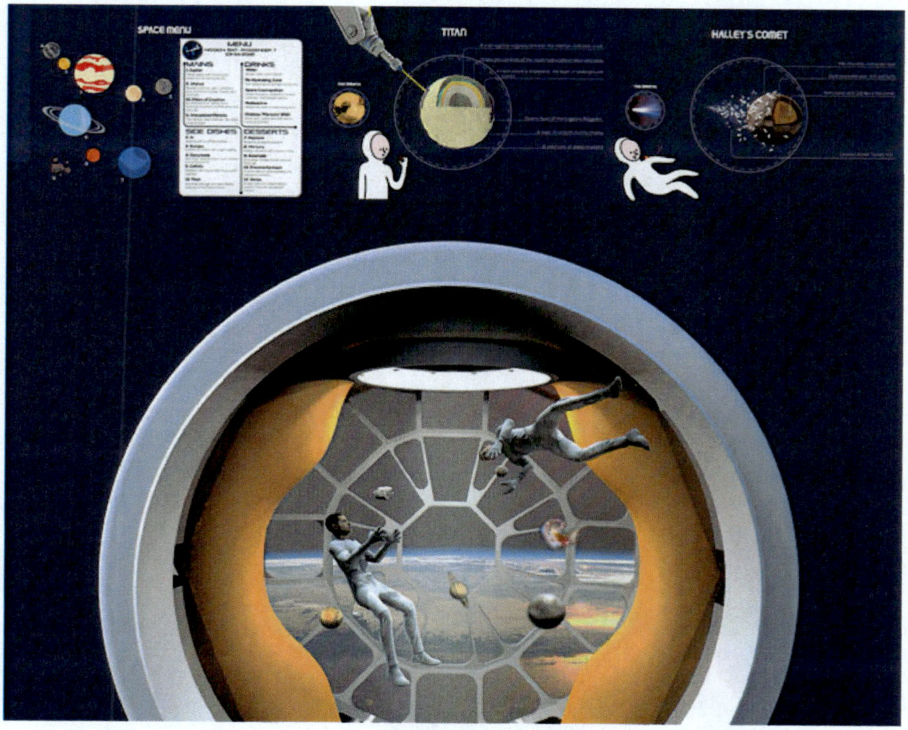

Fig. 4.18 In the project *Virgilio* the food is printed in 3D taking inspiration by the planets composition and translating liquid, gaseous and solid substances in edible ingredients for an immersive food experience in Space. Credits by *Space4InspirAction*, Politecnico di Milano

– *Interplanetary Missions*, including both the spaceships for the travel to another planets and lunar and martian bases; we had the chance to count for this third project of the support of Thales Alenia Space (TAS), the world's leading player in the construction of space habitats, to explore how to build a space habitation module, an orbital station with microgravity, a Moon or Mars base in reduced gravity.

The results of the 1st edition 2016–2017 of *Space4inspirAction (S4I)* have been presented during a public event at the *Planetarium Ulrico Hoepli* in Milan (Fig. 4.17), an exhibition for the *Long Night of the Stars*, an ESA event celebrating the 50th anniversary of ESOC, the European Space Operations Centre, in Darmstadt, Germany, and during the *European Researchers Night* in 2017 at the ESA Italian headquarters in Frascati, Rome.

In the next pages a short excursus of the projects to give an idea of the big variety of the ideas, all of them driven by the desire to make closer Space and Earth and increase the well-being of the astronauts producing spin-offs and spin-ins.

Fig. 4.19 Inspired by the lotus the space food is inside the flower-shaped packaging *Bloom* that opens its petals when heated (left) while astronauts use *Shake It* to heat and mix food (right). Credits by *Space4InspirAction*, Politecnico di Milano

4.4.2.1 Space Food

Considering food has a big importance for the health and the well-being of the people in general, and in particular for the astronauts who live in a confined environment without natural stimuli, the major part of the projects is about creating *a new immersive food experience in Space*, including also space tourism.

In *Virgilio* (Fig. 4.18), people go on these trips to experience what an astronaut experiences, microgravity, feeling weightless, eating in Space, where food is easily grown and processed on board of spaceships. By using an advanced form of cellular culture, we are able to bring just a few *stem cells* of different typologies of food, which we can, through a computer-controlled environment, grow when required, and through a laser system transform them into a printed food with shape, dimension and layers determined by planets, comets, asteroids or stars compositions, that are translated into ingredients. For example, asteroids have a frozen core but solid crust, so it could be translated into a solid cookie core with a chocolate crust flying through the module. In this way, you can learn about these space objects while eating them. The production chain explained earlier gives us the possibility to produce anything we want, when we want. An important factor to consider is the experience of eating this special food. That is why we also devoted some time to the module in which the space tourists will consume it. The module

Fig. 4.20 Thanks to *Leaf* system astronauts can cultivate, grow plants, carry out research and experiments, feed in a healthy way and maintain an emotional relationship with the Earth. Credits by *Space4InspirAction*, Politecnico di Milano

consists of a sphere (which recalls the food the travellers will eat), which allows people to experience microgravity to the fullest (as opposed to current volumes within the ISS). To help them move around, we designed two paths on the inside surfaces of the module, on which a nanotechnology adhesive helps them walk around the sphere and look around. Two larger windows fuse outer space with the confined volume of the module, to enhance the experience of floating outside the Earth's atmosphere. This three-dimensional walk is something that cannot be done on Earth and will add to the amazing experience. The food that has been produced is floating around inside this module, and the space tourists will have to grab it in order to eat it. Fortunately, they are not alone. A guide, *Virgilio*, is there to help them through this voyage, taking them through the space menu as an exploration of the universe, explaining them what they are eating and where they are located. In its entirety, this explorative concept looks at a new way to produce food, but more importantly, the way this food can carry a message and the valuable experience it creates for space tourists.

Coming back to astronauts, *Bloom* creates a poetic and seasonal experience taking advantage of the microgravity: inspired by the lotus in the pond and lanterns floating on the river, food is stored in flower-shaped packaging, which once heated, rise upwards opening the petals slowly until they form dishes that astronauts take and consume without the need for other tools. *Shake It* suggests astronauts use

Fig. 4.21 *Mexe* system consists of an EVA suit that increases the physical capabilities of the astronaut during exploration activities on a lunar ground supported by a multifunctional vehicle. Credits by *Space4InspirAction*, Politecnico di Milano

different gestures to shake this heater (Fig. 4.19) in order to have a real experience of making food like on Earth, using the pot to heat and mix food.

Moving from space food tools to systems to grow plants transforming the spaceship or a habitat into a self-sufficient organism, in terms of energy and renewable resources, the self-sustaining *Leaf* module is a concept designed to support space explorers during long trips, thanks to the growth cycle of plants on space stations or planetary surfaces that allows for the feeding of the crew and for scientific research. Starting from MELISSA, Micro Ecological Life Support System Alternative—a ESA project for a closed life support system aimed to produce food, water and oxygen for manned space missions increasing regenerative behaviours—*Leaf* module promotes in a whole process of farming, researching, eating and recycling, a strong interaction between the crew and the plants, creating an emotional connection with Earth and maintaining an environment surrounded by plants that provide explorers with air and a familiar habitat as well as a better nutrition (Fig. 4.20).

On board the International Space Station (ISS) astronauts always need to have their hands free to grab any tool in an emergency. The lunchtime is always fast and often takes place in a matter of minutes between a commitment to another. On the opposite side, in view of these new types of tourist travel, there is a need to rethink the food system, as for a tourist feeding time must be quality and at the same time a new experience. *Gulp* is composed of two main elements: a small cuff, always worn by the astronaut in the upper part of the forearm and an elastic band, which is inserted into the forearm only at the time of the meal. When the envelopes are inserted into the rails of the bracelet, thanks to sensors placed on the bracelet, the foods are recorded so you can track your meal to meal the very delicate astronaut

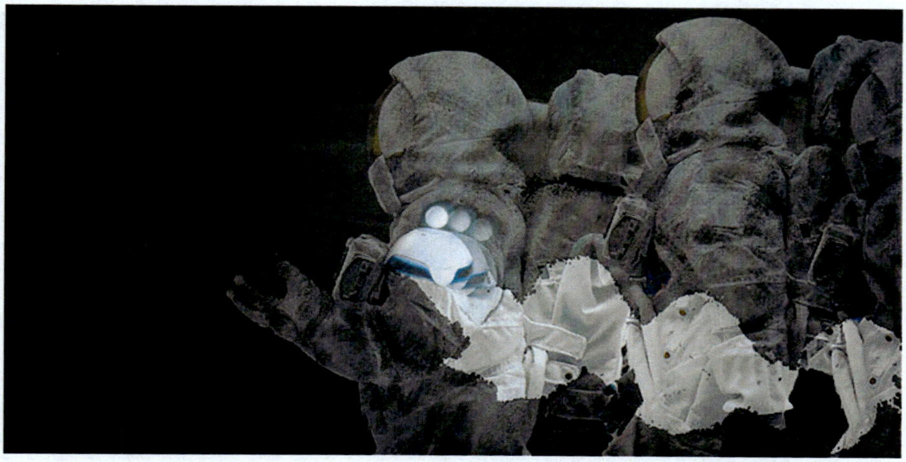

Fig. 4.22 *Robot Pet* cleans the lunar dust on the EVA suit using an electric field meanwhile supports astronauts in an interactive way facilitating the communication of the crew during lunar exploration missions. Credits by *Space4InspirAction*, Politecnico di Milano

diet. Thanks to its high practicality, it can also be used in sports and work extreme situations wearing the bracelet during cycling or climbing. A good example of spin-off of both technology and behaviours from Space to Earth.

4.4.2.2 Extra-Vehicular Activity (EVA) Suit

The EVA suit opens to many new interpretations considering that in the near future astronauts will pass in the same day through different kind of environments and gravity, orbiting around the Moon in microgravity, exploring its surface with 1/6 g and living the transition from inside the pressurized habitats to outside, floating in the vacuum or walking on the ground. At this purpose *Mexe* is a concept designed to facilitate the exploration and construction of settlements on the Moon; the system consists of an EVA suit that increases the physical capabilities of the astronaut supported by a multifunctional vehicle (Fig. 4.21). The suit has an additional feature, an internal exoskeleton that increases the astronaut's physical capabilities and reduces fatigue and stress from EVA. The exoskeleton system adapts to the astronaut's anatomy, identifying points of articulation that require support and maintain an individual profile of each user, learning and adapting to the requirements of the astronaut. To facilitate the development of all activities that are required, the EVA suit has been developed which keeps the astronauts in a controlled environment and provides the necessary resources for their survival, it also protects them from external threats such as radiation, meteorites, or debris.

A disruptive idea to eliminate the removed lunar dust from the EVA suit is to imagine the suit breath and through this movement reject debris. In addition, the outer layer of the breath clothes uses a special material that produces positive

Fig. 4.23 *Mars Transfer Vehicle* is designed to travel from Earth to Mars changing gradually the gravity to adapt the human being to the new environment. Credits by *Space4InspirAction*, Politecnico di Milano

charges in the Sun's radiation like *Sun shower*, thus repelling the same positively charged lunar dust. But it still cannot exclude the possibilities that some large particles dust cannot be repelled and still stick to the clothes. By realizing the movement of breathing to remove large particles dust, we use a particular material under the outer layer that has the characteristic like the sponge that can capture gas, store gas and release gas, when cleaning the EVA suit, and this layer will be filled with the gas and it has the trend to go out affected by pressure difference due to the vacuum environment of the Moon, thereby causing the clothes to expand while removing the gas. In the shape, we hope to strengthen the visual effect of *Breathing Suit* so that astronauts can clearly feel inhaling and exhaling of the clothes, so we use three-dimensional geometric appearance, where the clothes maintain some small folds in the normal state and expand instantly when removing dust. We suppose the using environment is that astronauts enter a preparatory cabin to make their EVA suits remove lunar dust, waiting for the cabin to finish collecting dust, then leave their EVA suits and go into the master cabin, where the cleaned clothes can be used again.

Another concept design, *Robot Pet*, is inspired by clean fish in the sea that helps to clean bacteria and other microbes parasitizing on the body of big fish. The main function of this concept design is cleaning the lunar dust on the clothes; meanwhile, it can be like a pet to accompany and help astronauts when they engage in the lunar exploration (Fig. 4.22). The principle of removing dust is also to use the electric field. *Robot Pet* can move on the astronauts with four wheels covered with the material nano-level bristle structure like gecko. It has an internal cavity with electric field to absorb dust and collect them in storage bags and each part of the object can be disassembled and cleaned for recycling use. In addition to the dust removal

Fig. 4.24 This *Mush Rooms* habitat system grows up thanks to a mycelium inside an inflatable structure that breaks when it has completed its growth and dying has crystallized. Credits by *Space4InspirAction*, Politecnico di Milano

function, we imagine it can also be effective as a *pet interaction* with the astronaut, which is realized primarily by voice and pet head effects. The head of the robot pet can shake as the music playing, it can change the colour, according to the astronauts in different states, and display astronauts' health status data and give warnings. It can also increase the experience of communication between astronauts looking at each other's pet activities to engage in dialogue.

Considering the complexity of the EVA suits and the low visibility, it is important to support astronauts with smart tools able to become additional prostheses to be used as extensions of the human body, like in this example. All the projects are based on real technology projecting new applications and disruptive ideas to increase astronauts' performances and well-being.

4.4.2.3 Interplanetary Missions

A long-duration spaceflight, a challenging environment in confined spaces, together with other people, travelling towards a new home. This is the chain of thought to start defining new concepts to go to Mars, for example, with the *Mars Transfer Vehicle*, designed to travel between Earth and Mars bringing people and research equipment (Fig. 4.23). Because long-time exposure to microgravity has severe physical consequences, we designed the spacecraft to have *artificial gravity*. The ring at the front of the craft, to which the habitation modules are also connected, rotates creating a centripetal force that equals that of the Earth gravitational field, helping to counteract those negative effects, and it also has a positive psychological effect to the occupants, being able to move like on Earth.

Fig. 4.25 The hexagonal section of the Gateway gives a new aesthetics with more perspective and space for storage including passage of cables and systems. Credits by *Space4InspirAction*, Politecnico di Milano

The centre of the spacecraft (cupola) does not rotate, so occupants can still occasionally experience microgravity. An interesting aspect of this project is that of the artificial gravity. After the modules have been connected to the spacecraft around the Earth in orbit, the ring will start rotating so to recreate the Earth gravitational field of 1 g. However, the gravitational field of Mars equals 1/3 g. So, as the journey through space proceeds, the rotational speed of the rings decreases (centripetal force is directly proportional to rotational speed, with fixed diameter) as to slowly go matching the gravitational field of Mars, 1/3 g. In this way, the transition is smooth for the settlers, and they can start training and getting used to the new gravity field. The journey can last from 5 months to as long as 1 year, so we developed the spacecraft to take advantage of all this time. On Earth, the settlers will receive basic training regarding space travel, spacecraft operations, social training and others. When they have finally arrived on the spacecraft, they will receive specific training for their activities on Mars supported by doctors, engineers, nutritionists, personal trainers, psychologists, scientists, etc. Within the ring, they can find various training areas to use. Laboratories, physical exercise, an area where the surface of Mars is recreated for movement exercises. The ring also contains the shared eating facilities as well as sleeping quarters for the crew.

Once arrived on Mars, we should have a colony ready to welcome us, like *Mush Rooms*, an independent construction system of modular cells for the interplanetary colonization (Fig. 4.24). We got inspired by the progressive building of a society in

Fig. 4.26 Interactive screens cover the racks instruments of Gateway to increase a visual order for a more efficient use of them both for working than resting activities. Credits by *Space4InspirAction*, Politecnico di Milano

a specific context. For this reason, we recreated this gradualness with the use of a material that has its own growing steps. Moreover, the idea of mushrooms was conceived not to *bring* a material from the Earth but *generate* it on the planet itself.

4.4.3 Space4InspirAction 2nd Edition 2018

The 2nd edition confirmed the presence of Thales Alenia Space—a joint venture between Thales 67% and Leonardo 33%—the world's leading player in the construction of housing modules for the ISS—who chose the collaboration with our *Space4InspirAction* Course to design the Gateway, the new cis-lunar station, on which all the space agencies are concentrated at this time.

The results have been presented at the Altec Thales Alenia Space headquarter in Turin during an event with an exhibition in which all the engineers working on space habitat topics took part and we gained a lot of satisfaction confronting with them on our project solutions. The main objective of the twelve projects presented was to improve the living and working conditions of the astronauts, focusing on the well-being and psychological and emotional aspects that, together with biomedical ones, are essential for the success of a mission: integrate augmented and immersive VR to allow a wider and varied perception of space; use water as a shield against

Fig. 4.27 The perception of the internal spaces of the Gateway are transformed by VR to divide rooms and recreate others artificially both collective for gathering then personal to gain more privacy. Credits by *Space4InspirAction*, Politecnico di Milano

cosmic radiation; introduce illumination systems that match the circadian rhythms; integrate sensory environments to be able to stimulate all the senses and innovative technologies that reduce noise on board. The work activities are facilitated by new equipment that transforms microgravity into an advantage, favouring fluctuations that are possible without weight, and much importance is given to entertainment, both personal and collective, promoting privacy as well as conviviality to strengthen the membership of a group.

Among the projects, there are proposals for space habitats that exceed the internal square section of the current spatial modules proposing hexagonal and octagonal sections useful for obtaining more space to be dedicated to areas of containment and passage of cables and systems, as well as being aesthetically more interesting because of the new dynamic visual perspective (Fig. 4.25). For the management of the storage, a new internal structure has been developed, using aluminium and fabric to make it lighter and customizable: the easy movements of the flexible structure make it easier to customize the containment. The panels have a pattern created to generate interactive experiences using a soft and technological material that contains lights for the lighting of the module in accordance with the rhythms of the Sun on Earth that mark the time of the astronauts on board.

In some projects, the internal volumes are completely redesigned and alternate large areas with foldable areas that are transformed according to the various needs of space; mobile and light structures, can be fixed at any point of the station to work or rest; interactive screens cover the instrumentation in order to achieve an overall visual order and a clearer perception of the whole environment (Fig. 4.26). *C-Touch* is a system based on the combination of fabric and projection that wants to

Fig. 4.28 *Core* shelter protects the Gateway by cosmic radiations thanks to water bags which can expand or compress according to protection and space needs. Credits by *Space4InspirAction*, Politecnico di Milano

give the space inside the Gateway module a good level of visual comfort to improve the psychological needs of the astronaut. The aim of the project is to create a visually clear and orderly environment that interacts with the astronaut and modifies the space according to the different times of the day. The mobile textile system covers the two longitudinal parts of the module, one dedicated to rest and the other to work. In the working part, the curtain can be raised and lowered by hand, during working hours the fabric can be closed to use the entire working space or it can be partially opened and become an extension of the interfaces for the astronaut.

The spaces are transformed also according to the activities that are carried out, and in these operations light and augmented VR help to divide the spaces and to recreate others artificially, creating environments clean and comfortable (Fig. 4.27). Many schemes of *Virtual Escape* are foreseen to change the perception of the spaces, both collective then individual, for example, the perception of privacy is also linked to the possibility of isolating oneself and immersing oneself in the home environment wearing a special helmet.

The examples are intended to give a general idea of all the projects developed, without going into details, because they could not be treated individually as case studies for lack of space. What is important here, is to give an idea of the diversity of the proposals that explore new configurations and space partitions to make the Gateway more comfortable, efficient, safe, and attractive for the crew.

The project *Core* revolves around the idea of protection: first of the crew members from the effects of radiation, and secondly, of their privacy in space. A modular network of shielding made with water inside integrates with the current water system in Space (Fig. 4.28). *Core* monitors and ensures water efficiency in

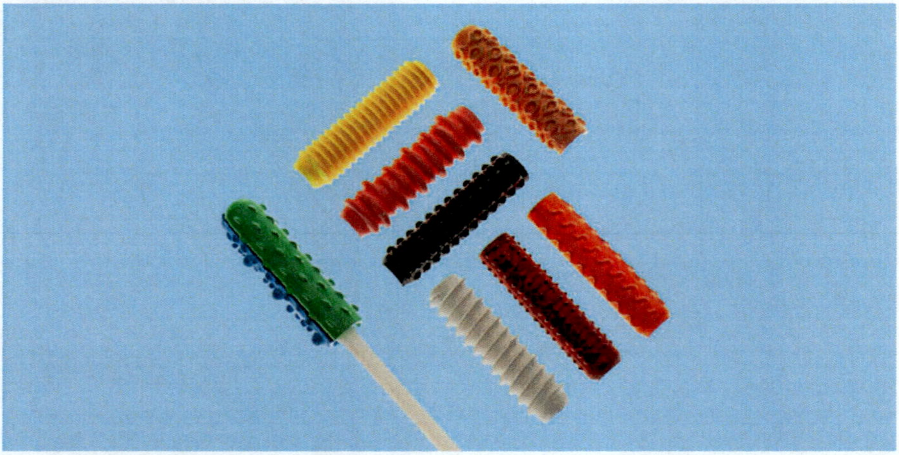

Fig. 4.29 In Space we can activate sensorial perception in different ways to contrast the lack of stimuli, as this food tools set able to stimulate the *Taste Buds* through different shape and textures. Credits by *Space4InspirAction*, Politecnico di Milano

Space and exploits the possibility of predicting solar storms to generate localized and targeted protections against galactic rays and cosmic radiations. The modular *water bags*, inspired by a bellows, are designed to expand and compress in order to provide astronauts with sufficient protection depending on the intensity of the storm. By providing localized protection, this shielding system will maximize the scarce amount of water on board the cis-lunar Gateway station through integration with the unused water, recycled or drinking water system. The amount of water needed to run the system is, therefore, what is currently provided in the module of the cis-lunar station Gateway, without the need for an increase. The modular and swivel interior has the aim of maximizing the perception of a private space within the common space: this new habitat, obtained through the rotation of a slice, creates a kind of personal refuge that offers comfort, relaxation and, above all, feeling of private space.

Considering that the Gateway will be smaller than the International Space Station (ISS), the need of privacy, already present on board the ISS, will increase. For this reason, alternative design solutions which could offer light ways to recreate private spaces, even if using VR, are desirable.

4.4.4 Space4InspirAction 3rd Edition 2019

The 3rd edition of the Course activated a collaboration with the Fashion Institute of Technology (FIT), New York, which design the packaging to enrich our products: we focused the topic on *Space Objects* inspired by Space scenarios and designed for

Fig. 4.30 When nostalgia increases in Space *SendSense* is a device that helps connecting people together by sending their sensation of touch to a loved one, who is far away. Credits by *Space4InspirAction*, Politecnico di Milano

astronauts with extensions to our daily life on Earth, like spin-offs of objects to be sold as merchandizing for *space enthusiasts* through the ESA online shops.

We presented the results to the specialized press projects with a detailed description of the idea behind the object, a professional quality video that showed those objects in an attractive way, and the possibility of touching the prototypes of the objects themselves, made with 3D printers and also equipped with a customized packaging. The twelve proposals considered the *physiological alterations* of the human body due to the microgravity conditions which lead to an overall reduction in sensory perceptions, such as, for example, the desensitization of the taste buds that leads to lack of appetite and, consequently, to incorrect eating habits, then also reflecting on the mental health of the astronauts. The projects aim was to increase the well-being by *activating the sensorial perceptions*, for example, with a set of tools that, recalling textures, shapes and colours of familiar foods, act on the taste buds of the astronauts, allowing them to stimulate and fully enjoy every type of food (Fig. 4.29). Moreover, once stimulated, the *Taste Buds* remain more sensitive for a certain period of time, before returning to their desensitization condition, allowing users to better enjoy all the solid foods they have. Finally, each tool goes to occupy half of the straw, the part in contact with the tongue, so it is also possible to combine two different ones and mix different sensations and shapes. This choice was made taking into consideration the subjectivity of the theme; each of us, in fact, reacts differently to flavours, based on the sensitivity of his taste buds, and in this way, he is free to try and combine the tools that allow him to amplify his favourite flavours. Although designed specifically for astronauts, this project could also be widely used in everyday life, starting from temporary conditions, for example, bad

Fig. 4.31 *Pare* brings the gestures of paring and enjoying fruits thanks to an edible packaging made of algae which contains a fruit powder that swells with water rehydrating in a star fruit like a lemon. Credits by *Space4InspirAction*, Politecnico di Milano

flu that can make the taste of food unpleasant, up to permanent physical problems, perhaps due to injuries or pre-existing conditions.

The sustainability principles which push us to imagine to re-use everything, including waste, overall on the International Space Station (ISS), inspired a system that allows sprouts to grow from the seeds in conditions of microgravity reusing the moisture collected in the towels after physical activities and showers. From the other side, we need emotional projects to help astronauts filling the gap with home, so why do not imagine using new tools to convey the sensation of touch millions of kilometres from Earth and give an astronaut, through a series of impulses, the warmth of an embrace? *SendSense* is a device that helps connecting people together by sending their sensation of touch to a loved one, who is far away (Fig. 4.30). Conceived for astronauts on the ISS to interact with their family on a deeper level than just voice chat, it is made of two parts: a core, in hard material, and a softer part grasping around the core. While the former is static, and mostly used as a charging and transmitting base, the latter is made to be wearable, leading to two different scenarios. The product has a twin placed on Earth and owned by the astronaut's family: the two work mutually both sending and receiving *touch messages* in the same way.

All projects are focused on emphasizing the importance of gestures, to design new ways of using objects and tools that could make the experiences more meaningful and profound. *Pare* brings the gestures of paring and enjoying fruits in an environment where, until today, is not possible to conserve fresh fruit for more than two weeks (Fig. 4.31). Inside an *edible packaging* made of algae there is a fruit powder which swells with water rehydrating in a star fruit like a lemon. Pare saves

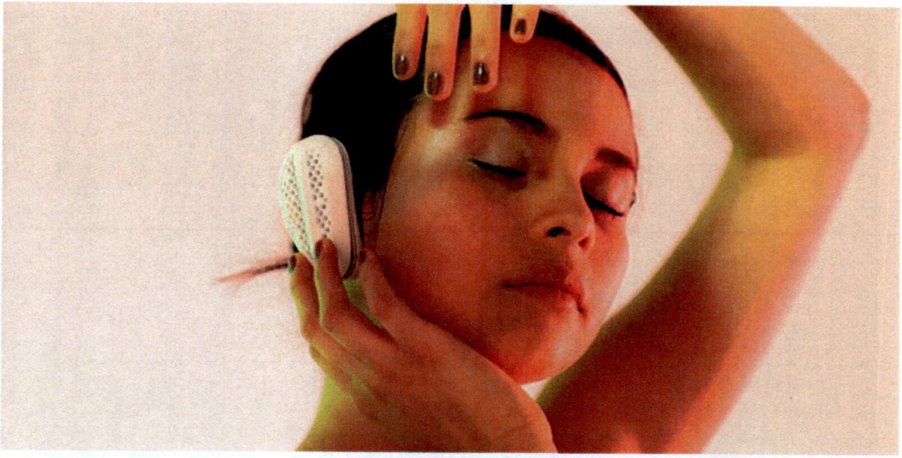

Fig. 4.32 *Onanze* reduce the noise on board the ISS erasing the frequencies and allowing the astronauts to choose their own background sound and maintain a good balance increasing the well-being. Credits by *Space4InspirAction*, Politecnico di Milano

space and weight while being transported, and more importantly, there is no residual waste to be stored and discarded afterward.

Also, body care and personal hygiene are linked to the senses as they are needed to increase well-being. Very simple and trivial actions on Earth, such as cutting your nails or washing your hair, are very difficult in microgravity and in a confined environment: and some solutions propose simple and functional alternative systems—all based on the idea of collecting and containing waste and water to avoid that floating will clog air filters and on board instrumentation—which can also turn into terrestrial spin-offs. The common intent of projects dedicated to personal hygiene is to transform these activities that in orbit are very complex in pleasant moments avoiding the stress of performance in microgravity.

Another occasion of stress on board the International Space Station (ISS) is also given by the continuous background noise. In long-term space missions, astronauts can feel stress and discomfort not just because of the pressing daily work schedule, but also because of the environment of the ISS. The ventilation system is composed of several numbers of fans, which end up with a continuous, loud background noise. Aim of *Onanze* is to solve this issue by reducing the fans' noise in a passive way: the mechanical soundwaves are forced to bounce in wavy canals, so the specific frequency of the disturbing noise is erased.

The soft part of the product is in contact with the skin and is tailor-made according to one's specific ears shape (Fig. 4.32). Along with its outer profile, there are three little balls placed in an exact position following reflexology principles. In addition, this object has a microphone connected to the ground control in order to monitoring astronauts acoustic stress level. As a spin-off on Earth *Onanze* may be used in noisy working environments, its uniqueness lays in the perfect adherence to

Fig. 4.33 *Fitness in Microgravity*, from the lens of Design, means also to perform other sports, as ski, in an immersive virtual panorama that recall our wonderful mountains on Earth. Screenshot taken by the author during *Space4InspirAction* 4th edition, Politecnico di Milano

the ears, which does not require any wires, nor headbands. It has not been designed to isolate, likewise usual headphones, but to enjoy your own sound environment.

There is not enough space to describe all twelve projects, each with its own specificities, what is important here, is to give an idea of the diversity of the approaches that look at life in Space and the habits of astronauts as an opportunity to improve some performances, that in the absence of gravity become more difficult.

Focusing on human factors as well as physiological and psychological behaviours in relation to objects and the environment, we can increase the comfort and efficiency of performances of astronauts on board the ISS, making their activities easier and more pleasant, developing the ability to observe and predict new uses and gestures: the *Use and Gesture Design (UGD)* methodology is particularly useful when we confront with small and handled objects which can solve even small problems but useful to increase the general well-being.

Fig. 4.34 *Lighting in Space*, from the lens of Design, means to change the perception of the environment using lights to divide spaces or create volumes for gathering or privacy needs. Screenshot taken by the author during *Space4InspirAction* 4th edition, Politecnico di Milano

4.4.5 Space4InspirAction 4th Edition 2020

The 4th edition of the Course *Space4InspirAction*, that is going to the end while I'm writing this book, was characterized by experience of *distance learning* due to the Covid-19 pandemic which forced all of us to stay confined at home increasing the need of privacy like it happens on board the ISS (Clearwater 1990). This new condition allowed us to choose two themes for the projects that could be lived in the personal experience of the students in a confined environment: being in isolation as astronauts they could better understand their needs and behaviours and find innovative solutions to increase the comfort and the well-being in Space (Dominoni and Quaquaro 2020).

We foresaw two projects, one oriented to designing in-depth *technical machines and tools* following strict requirements for confined and microgravity environments, one focussed on creativity and the power of Space to inspire the common imagination, looking at science fiction and *Space Age* as examples to define new *emotional scenarios*.

The first brief (Space Action), developed during the month of April 2020, was dedicated to *Design for Space*, with the topic *Fitness in Microgravity* focused on the sensorial and practical aspects of fitness activities in Space, in absence or reduced gravity and in confined environment. The aim was to re-think new machines and tools for the physical activity and other terrestrial sports to practice, as skiing (Fig. 4.33) or swimming, on board the International Space Station (ISS),

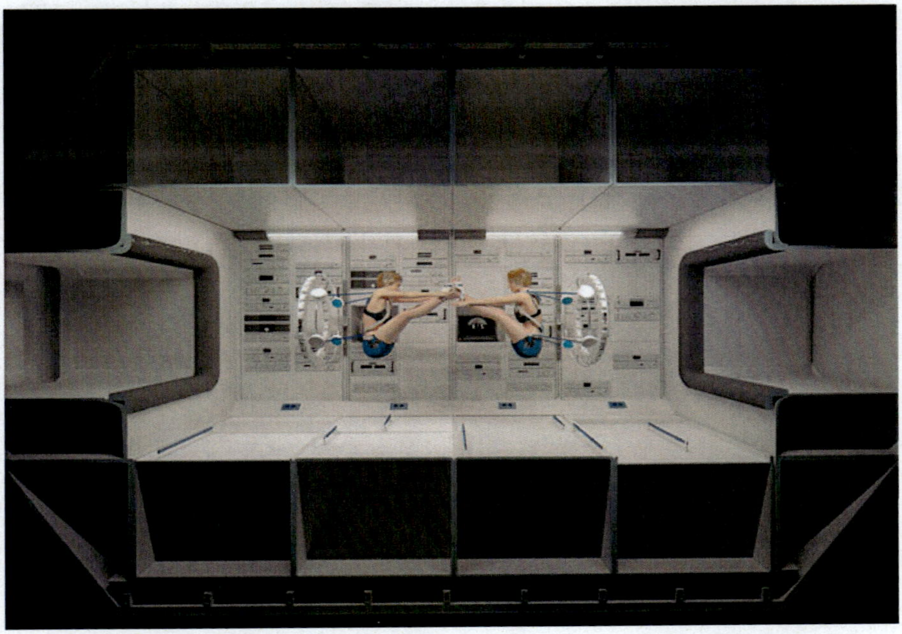

Fig. 4.35 What happen if astronauts perform acrobatic exercises in microgravity? Credits by *Space4InspirAction*, Politecnico di Milano

but also imagine how astronauts could make the same activity inside the future cis-lunar orbital station called *Gateway* and/or on the lunar surface, thinking about the life inside the new human bases on the Moon. We wanted to increase the comfort and the performance of astronauts, reduce weight and volume, but also, transform boring physical exercises in entertainment and increase the quality of crew relationships through the *gamification* of fitness activities. As usual, we believe that in a Course/Laboratory of Design it is very important to involve a real company with a specific know-how on our topic: so, for this first project we asked to TechnoGym, a leader in the world in designing fitness machines for *wellness oriented* customers.

The second brief (Space Inspiration), developed during the month of May 2020, was dedicated to explore the most emotional aspects of Design, analyzing the suggestions that Space has exerted and exerts on the terrestrial life of every day, to get to define terrestrial *Space inspired* products. The aim was to focus on the power of Space scenario to inspire the common imagination through science fiction, literature, movies, music, but also environments, architecture, fashion, art and design. Students were asked to choose a sci-fi movie and imagine *Lighting in Space* defined by style, shapes and colours, in which the role of the *light has the great value to change the perception of the environments* (Fig. 4.34). For this project, we involved

Fig. 4.36 Technology and poetry interact with lighting to obtain *komorebi* effects and re-create natural circadian rhythms. Credits by *Space4InspirAction*, Politecnico di Milano

the company Foscarini, an Italian excellence in the field of lighting design, overall regarding decorative lamps, very known worldwide.

Choosing this two projects we wanted to face students with the human factors aspects, putting *the user at the centre*: we asked them to deepen into the technical details of the fitness machines currently used on the ISS, considering at the same time the physiological alteration suffered by the human body in microgravity, and underlining the essential impact of the *Usage and Gesture Design (UGD)* methodology, which we have created to support Design for Space. From the other hand, we intended to stimulate the imaginary of the students pushing them to go beyond the boundaries of the project, to let the fantasy free, to make different associations, no strongly related with the reality, and explore the Space scenarios and the sci-fi movies inventions to generate powerful objects, overall from the *visual point of view*, like it happened during the Space Age. The results we gained were very interesting and surprising for both the themes.

In the 1st project of *Fitness in Microgravity* students went in depth into technical details in order to increase the performances of the astronauts during fitness activities introducing new topics connected with *entertainment*—as the possibility to *perform other sports*, like swimming, skiing, or climbing, according with the microgravity conditions—and with the need to *relax and meditate*—imagining immersive bubbles to perform pilates and yoga—including also the possibility to strengthen the team spirit with exercises to do in pairs and groups. In this example, acrobats inspired the project and the trick was to translate choreography and movements in a version for Space, where microgravity makes possible exercises unthinkable on Earth, but at the same time, complicates the ability of astronauts to balance and maintain position, overall if exercises are thought to be performed as well as individually, in couple and in groups (Fig. 4.35).

In the 2nd project of *Lighting in Space,* they explored the various possibilities of the light to change the perception of the environment, making spaces larger or bigger according to the different needs of the astronauts, using the power of shadows to project natural effects like the poetic *komorebi* (Fig. 4.36), the light that filters through the leaves of the trees, enhancing the comfort zone of the private spaces giving the feeling of a warm and cosy environment with soft light, or again, increasing the well-being of the crew working on the power of light to re-create natural circadian rhythms. In particular, light projections are used to depict the reflections by hypothetical windows, giving the illusion of being in a room with openings. The lighting system is thought to be easily integrated into the current lighting system present on board. An OLED screen is placed inside a cylindrical mask with cut-outs. The mask can rotate and it comes with the silhouettes of two windows, being customizable. The OLEDs mimic the temperature and the intensity of natural light in two different modes, automatic and weather, which can be set using a small switch recreating a sunny, cloudy and rainy day atmosphere depending on your choice.

4.5 Fashion in Orbit Versus Space Fashion Design[10]

The extraordinary results of the project *Couture in Orbit* presented in London have since generated over twenty million media feedback responses and this inspired us, along with the European Space Agency (ESA), to create a specific Higher Education Course called *Fashion in Orbit* at POLI.design (the consortium of Politecnico di Milano) in February 2017 and based on the same idea of a *capsule collection* of garments to be worn on Earth that integrates space technologies and innovative materials. *Fashion in Orbit* was not a curricular Course but was open to young professionals who want to develop innovative products in fashion inspired by Space and technology research in order to find new applications and spin-offs for materials and textiles. The 2nd edition 2018 we transformed the Higher Education Course *Fashion in Orbit* in an academy contest called *Space Fashion Design* that expands the project field including fashion as well as product design and accessories, always driven by space technology—to follow the fast changes of these two worlds that increasingly overlap and lose their original boundaries by embracing

[10]*Fashion in Orbit* and *Space Fashion Design* are two projects of Higher Education Courses created and directed by Annalisa Dominoni and Benedetto Quaquaro at POLI.design (the consortium of Politecnico di Milano) with the support of the European Space Agency (ESA).

more fluid dimensions of project and technology—and transforms the format into a competition on invitation to young professionals, with expertise in Space Design, tailored to the needs of the reference company with which previously we built the brief together with ESA. The aim is always the same of the 1st edition 2017: to develop ideas of innovative products in the fields of fashion and design and find new applications and spin-off for materials and fabrics, inspired by space technological research.

4.5.1 Fashion in Orbit

The *Fashion in Orbit* Higher Education Course involved ESA and specific companies of technology focussed on sensors system and data collection together with many other companies coming from smart textile and luxury. It has been led from 30th January to 17 February 2017 for three weeks, eight hours per day.

Colmar, an Italian brand of sportswear, was the main sponsor, while the other companies—Alcantara, Caimi Brevetti, Coats, Limonta, Luxury Jersey, Omniapiega, Sitip, Soliani, Thermore—supplied materials, fabrics, accessories and fabric processing systems to realize the capsule collection. The aim of the Course was to create a capsule collection, using Space as a strong propulsive rocket to enhance imagination, and space technologies a starting point. We had the great luck to manage a very high level and composite resources to work with, having on the table fantastic fabrics, accessories and technologies to manage. They were so different from each other, as the composition of the international group made by students and tutors with different nationalities and background, and this variety has been a richness. Our first consideration is that the capsule collection of *Fashion in Orbit* is strongly influenced by this condition of multiplicity as our *universe* is made by many galaxies. Different flavours and different situation are floating and relating the one which the other with a result of a cosmic equilibrium. Second consideration is about the specificity and the use of technologies. We declared in our *manifesto* to use Space and Space technologies as inspiration. It has been clearly understood by students: they tried to get a strong suggestion by using different materials looking at them a fantasy enhancer. A fabric which could make a fantastic job for a specific use, may suggest an emotion or a poetic thought. Being *Fashion in Orbit* a conceptual work, we decide to declare the results of using sensor technologies by showing luminescent light signal, thinking of an implementation which will come. Third, consideration is about Space, Earth and visual topic. Fashion in the age of technology makes us understand better how we can be more conscious of what the transformation technology brings. Space innovations will have a strong influence on how people behave and perform on Earth. The *Fashion in Orbit* collection takes inspiration by Space and it has been developed to have a strong influence of terrestrial look. The visual aspect has a dominant part in the developing process,

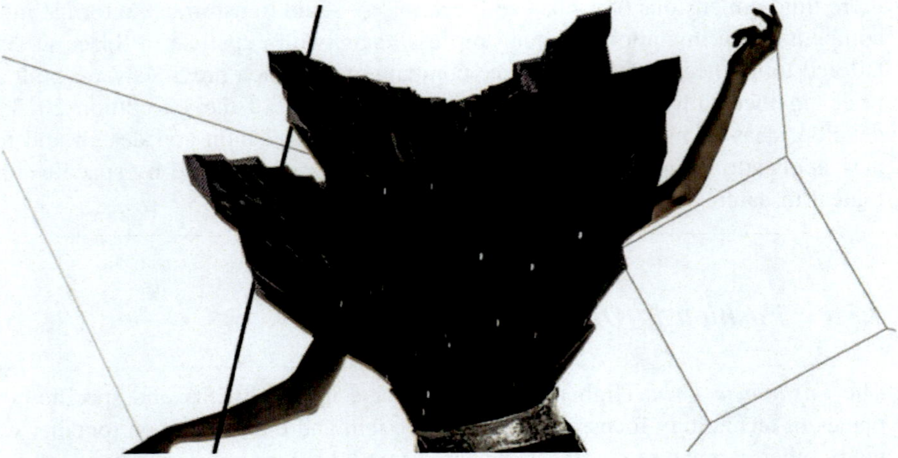

Fig. 4.37 The experimentation of different ways to use 3D fabrics together with the integration of bio-sensors and light signals produced a capsule collection characterized by a strong visual impact. Credits by *Fashion in Orbit*, POLI.design, Politecnico di Milano

with the goal to reach high impact images, typical of the fashion system. *Fashion in Orbit*'s pillars included fashion, technology and Space inspiration. We presented the capsule collection realized, thanks to the support of the European Space Agency (ESA) and the technical materials and fabrics given by the companies involved, during an event at the School of Design, Politecnico di Milano, which included the exhibition of all the space capsule collection prototypes. For the occasion, we realized a portfolio which collected all the works to offer to the companies. During three intense weeks, we led a group of international students working hardly approaching Space to Earth. The aim of *Fashion in Orbit* was to explore *new visions of fashion in the age of technology* taking inspiration by Space: upside down, floating, emotional state, connection, transition, personal space were some of the keywords we chose together with the group of students to link all projects in one capsule collection, imagining life in Space and the extraordinary experiences as weightless looking at the Earth from another point of view. All the conceptual works represented different identities and backgrounds and shared an intimate relationship with the cosmos, which is expressed through clothing with new languages full of emotion, feeling and poetry mixing fabrics and materials in unusual ways. The interactions between the human being, the body and the environment were explored and reinvented suggesting new needs and scenarios in which all the senses participate in the dialogue with the garment by activating bio-sensors, luminescent signals and mechanical movement to make explicit our intentions (Fig. 4.37). We think these great achievements in such a short time could be an inspiration for all the companies which participated in *Fashion in Orbit*. Our wish is that some of the many ideas emerged during the development of the Course could

Fig. 4.38 Space materials and technologies provided by the European Space Agency (ESA) and the private companies involved have inspired the capsule collection designed to be worn on Earth. Credits by *Fashion in Orbit*, POLI.design, Politecnico di Milano

generate innovation and become new products in the real market thanks to the involvement of one or more companies that participated interested to the outputs.

4.5.1.1 Learning Goals

The Course was tailor-made for those who wanted to explore fashion in the age of technology and better understand how to become a more aware protagonist of this transformation process, knowing that Space innovations will have more and more influence in the near future on behaviours and people's performance. It offered a unique opportunity to build transversal relationships with multidisciplinary teams of experts from the most technologically advanced sector in the world, as Space is, and to explore how research conducted in Space can inspire fashion and create new opportunities for social and business growth. Participants had the opportunity to create a capsule collection characterized by a high level of innovation, thanks to the collaboration with leading industrial partners in the fields of smart textiles, technologies and wearable, and to present it through a professional portfolio with a strong and recognizable image.

4.5.1.2 Syllabus

The Course *Fashion in Orbit* consisted of two main modules, the first made of theoretical content, the second about Design and visioning to support the development of the capsule collection and the realization of clothing and accessories. The first module was addressed to professionals and companies operating in the fashion field and who intend to get in touch with ESA and the most qualified realities that make innovation and provide the results of their research to inspire new products. Lectures were led by scientists and researchers from ESA who presented space technologies, including specific case studies in fashion, and indicated the most innovative applications to explore possibilities to create a capsule collection according to the brief given by the main sponsor, in agreement with the other companies involved. The second module concerned the development of the capsule collection thanks to a highly qualified tailor's workshop supervised by us together with ESA and the other companies. We started with the transposition of the concepts in paper, then in canvas, then the realization of the garments using fabrics, technologies and materials provided by the technical sponsors going in depth into detail to design the executives drawings of the various projects (Fig. 4.38). Once the collection was complete, participants were involved into a *co-design* experience of the fashion photo shooting at Immagine Lab of Politecnico di Milano which integrated the project of mood, lights and the overall image, useful for the creation of the digital portfolio of the whole collection, tackling the various aspects of communication design with the support of external professionals.

4.5.1.3 Design Methodology and Learning Outcomes

The capsule collection was developed on the basis of a precise brief that took inspiration from space technologies among those presented in the first part of the course. Participants worked both in classroom, for the presentations of the companies' products and technologies, and in the Fashion Lab of Politecnico di Milano, with innovative textile materials reproducing the real dynamics of a fashion brand, from the presentation of the product to the communication system.

Crossing different knowledge we can achieve innovative ideas: taking inspiration by Henri Poincaré definition of creativity (Dominoni 2008), which is the ability to combine existing elements in new combinations, we believe involving in the same project many companies with different background, technologies and products was a good methodology to generate innovation. In this Course, it happened to be a surprise by the ideas emerged during the gathering with all the companies together with ESA, thanks to the crossover of know-how and experiences.

Participants acquired design and vision skills useful to collaborate with design companies and studios specialized in fashion design and innovation-oriented products. They practiced all the phases of the concept development for a new product of fashion, from the first ideas to the realization of the prototype, experimenting new material behaviours and technology applications. Also, they learnt how to create an innovative portfolio to present their work and how to valorize products through digital photo using light as a value to underline the principal aspect of the project.

4.5.2 Space Fashion Design

The 2nd edition 2018, compared to the previous *Fashion in Orbit*, extended the field of application to product design, in addition to fashion design, to follow the fast changes of these two worlds that increasingly overlap and lose their original boundaries by embracing more fluid dimensions of project and technology. It has been led from 5 to 16 February 2018 for two weeks, eight hours per day.

The nature of the Higher Education Course was changed in a more dynamic formula to allow us more freedom from an education programme, which presuppose a part of lectures and researches to introduce the topic of the project, to a contest on invitation for young professionals, with some expertise in Design for Space. Promoting this change of structure, we were able to open to a main sponsor and responding more efficiently to the brief we built together. Because of the appreciation of the results of *Fashion in Orbit* demonstrated by the companies we involved, it happened that the main sponsor of the first edition, the Italian sportswear company Colmar, decided to sponsor the *Space Fashion Design* contest.

The European Space Agency (ESA) supported *Space Fashion Design* Competition Workshop for the integration of technology through the programmes Technology Transfer Program (TTP) and Space Solutions—responsible for Italy Rina Consulting—and companies that already collaborate in Space projects, like Bercella Compositi, and Ribes Tech, a start-up supported by the Istituto Italiano di Tecnologia (IIT) and Politecnico di Milano in collaboration with Omet developing printed technologies, as photovoltaic cells and circuits. We defined a precise brief tailored on needs and interests of the company in connection with ESA Technology Transfer Network—Rina Consulting. The focus was on new concepts of ski garments taking inspiration by space technology with the aim to find new applications to increase the clothing performances. The project involves the development of wearable and integrable product ideas in sportswear—especially for skiing, but also extendable to other sports and sportswear in general—which were able to generate energy thanks to the developments of flexible photovoltaic technology by Enecom. RINA Consulting is also committed to carrying out a pre-feasibility study with the participants for the integration and operation of space technology that allowed to have the useful elements to make a strategic choice among the proposals presented. Parallel to the innovative technological solutions will be given equal attention to the style of garments and accessories that will have to align with the DNA of Colmar to be recognizable. Keywords: simplicity, linearity, elegance, with a strong attention to detail.

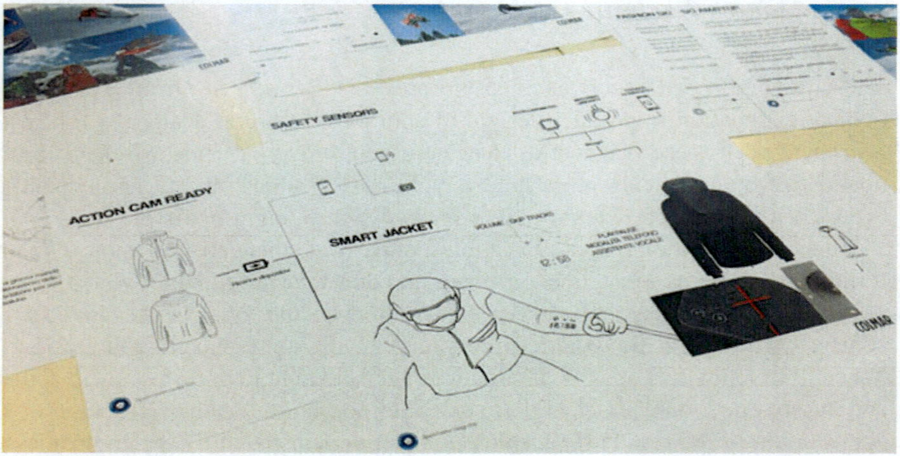

Fig. 4.39 Preliminary visualizations of concept design for a smart sky jacket which integrates space technologies. Credits by *Space Fashion Design*, POLI.design, Politecnico di Milano

4.5.2.1 Learning Goals

Participants had the opportunity to work together with ESA experts and big companies of fashion design like Colmar learning how to develop a whole project, from the first ideas to a detailed 3D concept, in which technologies and materials are defined together with fabrics and colour palette. *Space Fashion Design* offered a unique opportunity to build transversal relationships with experts and to explore how research conducted in Space can inspire fashion and create new business models created on innovative applications.

4.5.2.2 Syllabus

The 2nd edition was built on a competition workshop of two weeks full time (eight hours a day) structured as a design studio, with a group of young professionals and researchers led by us, with the partnership of ESA for the space technology inspiration, and with the support of Colmar to allow us to deepen the know-how of the company in terms of vision, strategy and products and understand better the brief's inputs. The concept was focused on the design of clothing, accessories and innovative products for ski field according to the needs of the company. Within the competition workshop, a session was devoted to the development of practical and strategic skills for the final presentation of the projects.

The outputs were divided into four distinct parts:

1. *Preliminary Research*, aimed at identifying product innovations under development and those on the market, and SWOT analysis, to assess Strenghts, Weaknesses, Opportunities and Threats;

Fig. 4.40 Space technologies as *Cooling Technology* and auxetics fabrics make clothing more breathable. Credits by *Space Fashion Design*, POLI.design, Politecnico di Milano

2. *Moodboard*, with indications of context, use, colours and materials;
3. *Concept Design*, descriptive of the general idea, visualized through sketches and renders (Fig. 4.39);
4. *Pre-Feasibility Study* of ideas, by Rina Consulting, which allowed us to gather useful elements to facilitate a strategic choice among the proposals submitted for possible future developments.

4.5.2.3 Design Methodology and Learning Outcomes

In this competition workshop, we applied the five principal methodologies that are the pillars of our design research and practice, plus other two specificities for Space Design discipline:

1. Design Thinking
2. User-Centered Design (UCD)
3. User Experience Design (UXD)
4. Design to Recicle + Design to Cost
5. Usage & Gesture Design (UGD)
6. Space Earth Design (SED).

The first four methodologies belong to Design discipline and focuss on the user needs with the aim to design a complete experience made of smart products, able to integrate technology and increase the performance while taking into account a sustainable design process. The last two are *dedicated* Space Design's methodologies, considered fundamental to design a new product imagining since the beginning of the development of the idea gestures and movements (UGD) how the astronauts could be in microgravity with it (how to use it, in which way, with which gestures, etc.). Making this effort of imagination we have an expansion of the context with more elements to integrate into the project and it is easier to create connection between living in Space and on Earth (SED).

One of the best skill participants learn is to make research activity for industry comparing information collected with the goal to achieve, moving in between intuition and feasibility, imagining new possibilities to use innovative technologies according to people's needs, for example, integrating ESA *Cooling Technology* for thermoregulation, or Electro Osmosis Process to extract sweat while skiing (Fig. 4.40). Space technologies by ESA are a strong motivational input that push designers and companies to find alternative and creative solutions for terrestrial projects starting from a perspective completely unknown. Participants experiment the importance to present their ideas with concept, images and words that are able to catch the attention of the companies' leaders, and they practice the power of story-telling to increase the value of the projects.

4.5.2.4 Projects Visions

The various proposals which emerged by the competition workshop provided to enhance the research prospective of Colmar with different innovation technology suggestions with the possibility to use more than one technology solutions, combining them. In order to help *thermoregulation* of the ski jacket one proposal designed a system of cracks in the padding, able to be closed and opened manually by the user, to second of the overhead-heating inside it. Lining of the jacket with cracks be closed and opened depending on the overheating inside, encouraging perspiration. The implementation is mechanical and manual. The first mode provides a layer of cotton wool lined, cut and sewn to ribbons. Pulling the strips, the flaps connected to the latter fold on themselves leaving free the holes of perspiration. The second modality previews the coupling of two complementary layers of perforated padding and overlap in two different positions. One of the holes match allowing the passage of air, the other two surfaces are completed by closing the openings. The position of the two layers is held by tapes.

Another suggestion introduces the use of the present space technology in the astronaut' suits: a system of pipes and air intakes manages to maintain the internal temperature of the stable ski jacket. Inspired by the thermoregulation technology applied to NASA's LCGV Suit, we thought to insert six tubes inside the Colmar ski jacket, which through openings on the arm, exploit the passage of air to cool the back, then expell the overheated air through an opening on the back, where it creates even an air depression. Also, exploiting the intelligent interweaving of *auxetic* fabrics to make the clothing more breathable and reduce sweating during stress. This concept is based on *auxetic* properties that can be obtained by applying certain cuts to the elastic tissues. These openings allow you to ventilate the interior of the garment ensuring a situation of well-being. The opening of the cracks is operated by some cables in Kevlar inserted inside the jacket to ensure breathability and waterproofness of the garment, thanks to an innovative fabric based on reverse osmosis. Heating the user using a hi-tech fabric will be able to transform electricity in heat even in extreme environments based on a technology developed by HydroBot, and this type of fabric allows to extract moisture through osmosis using electric current.

4.5.3 Results Dissemination and Exploitation

The results of *Fashion in Orbit* and *Space Fashion Design* were spread through the most important newspaper and magazines crossing the fields of science and fashion and we received many invitations to present our project during international events, exhibitions, symposia and scientific TV programmes among which

- Fashion in Orbit, special guest at the exhibition organized by ESA and DRL called *Living in Space*, Techtextil, Frankfurt 2017;
- Fashion in Orbit, event-talk at *Il Bello dell'Italia*, organized by the newspaper Corriere della Sera, Mantova 2017;
- Space Fashion & Science, TV program *Passi di Scienza*, Milano Design, RAI Scuola 2017.

4.6 Moony. A Lunar Base in Lava-Tubes for Igluna ESA_Lab@

We are witnessing an acceleration in the resumption of space exploration supported by cutting-edge technological research that will make possible new human missions to the Moon. Space agencies and industries predict the first permanent settlements on our satellite in early 2030.

The main aims of the researchers will be oriented to the deepening of the planetary sciences, like the origin of the solar system, but also the exploration and the finding of resources useful for the future expeditions towards Mars. Human permanence on the Moon will also give the opportunity to test the physiological and psychological effects that may emerge over a prolonged period of time in confined environments, subject to very powerful solar radiation, and in reduced gravity.

The private sector is also moving to make it possible i/n the near future to travel interplanetary with crews who will no longer be trained only by astronauts and scientists, but by tourists and travellers who want to have all the comforts possible to live in a pleasant environment. That's why Design for Space is crucial. As you can imagine the living conditions above and below the lunar surface are extreme for humans and require habitats designed to ensure their survival. In addition to confinement, due to forced cohabitation in indoor environments and without natural environmental stimuli, it is the distance from the Earth to emphasize the sense of isolation. Gravity reduced to 1/6, strong solar radiation, exposure to frequent impacts of meteorites and the extreme temperature range are some of the obstacles that man will have to overcome to ensure a continuous stay on the Moon.

A year ago, the astronaut Samantha Cristoforetti suggested my name to the Swiss Space Center as an expert in Space Architecture and Design because it was being born a collaborative project gathering university student teams from all the Europe around the theme of space exploration. *Igluna*, the pilot project of Esa_Lab@ coordinated by the Swiss Space Center and supported by the European Space Agency (ESA), was presented as an inter-disciplinary platform to create a space habitat to demonstrate how to sustain life in an extreme environment using space technologies for disruptive innovation and cross-fertilization. Three values support this project. First, to improve cooperation across Europe by building an international platform and collaboration between different disciplinary perspectives; second, to create an educational environment that is also able to inspire a potential future generation of space experts; third, to analyze the future development of space technologies being tested to discover possible new perspectives for space exploration. Teams of students from all over Europe, supervised and supported by their academic institutions, were called upon to develop a series of modular demonstrators inspired by the context of a lunar habitat. Together with students, institutions and professors, the aim of the Swiss Space Center was to try to involve another important actor: the local industry related to the territory of each university in Europe, to make all the projects feasible.

Creating a permanent habitat on the Moon, precisely under its surface, in the *lava tubes*, is a challenging project at this moment. I was invited, together with Benedetto Quaquaro, to participate in the first explorative meeting at the Swiss Space Center headquarter in Zurich. In that occasion, we presented our Course of Space Design *Space4inspirAction* and our expertise as space architects and designers to give shape to the future habitats looking at the needs of people and their behaviours, with the aim to increase the well-being of life in Space. As usual, it was not easy for us to be understood by the majority of engineers who sit at the table and explain them the role of Design for Space, but fortunately showing them examples of our works, the topics we promote became clearer.

Above all, it was clear that the Design discipline, taking care of the entire environment, becomes the coordinator of all the projects of components and tools developed by other universities; we had the task of integrating their projects in the best possible way so that the final result was coherent, functional and organic. We should have built the lunar base in which the other projects would have found their place, and we would also be charged with the exhibition of the final results. So, we decided to accept the assignment of supervisors and involve two of our best students to participate in the 1st edition of the *Igluna* project.

We were conscious that Design for Space can be of great help to suggest an approach more adherent to the needs of the human being in Space through the application of techniques and methodologies aimed at identifying security requirements, functionality, usability and pleasantness. Using Design, we have the advantage to increase the quality of life, not only acting with psycho-physiological factors, but above all with emotional ones: the greatest challenge is to create a space habitable system that is able to behave like a *living organism*, in perfect autonomy with respect to the earthly supply, especially in terms of non-renewable resources and logistical support. We realized also that our participation to *Igluna* project could be a good occasion to demonstrate to space scientific community the importance of Design approach and thought in all the phases of the development of space programmes and projects, from the beginning of the first ideas to the realization of the demonstrators.

4.6.1 The Roadmap of Igluna 1st Edition 2019

The *Igluna* roadmap started in September 2018 and continued until June 2019, when the project culminated in the *Field Campaign*, an exhibition of two weeks, which took place at the Matterhorn Glacier in Zermatt, Switzerland.

Such an extreme environment was the site for the demonstrator to be built, in order to test the challenge of designing with high constraints. Given the nature of ESA_Lab@, the unusual situation arises that the top-level design, e.g. the architectural structure was developed simultaneously to the subsystems. This required an even more pro-active and intensive exchange of information and tracking of interfaces. Besides, the cavity imposed some constraints: water (100.- CHF/m^3, only canisters), power and space are limited and the elevator to the cavity reduces the maximal size and weight of an indivisibly transportable piece (similar to the payload limitations of a launch vehicle). The habitat, which was built inside the glacier cave in Zermatt, was also accessible by tourists and media.

The event was split into two areas, one in the glacier, with the actual demonstrator and a cinema lounge, available for the projection of videos, and the second one in the town, at *Vernissage Art Gallery*, with an exhibition collecting visuals and explanations from some of the projects, not involved in any experimentation.

4.6.2 Multidisciplinary Teams: Activities and Methodologies

The 1st edition of *Igluna* 2019, brought together 19 teams from 13 universities in 9 different European countries. The multidisciplinary nature of *Igluna* project covers a full range of topics, going from the habitat conception and construction to life support systems, communication, and navigation, power management, as well as science and human well-being, and the last one, my favourite: if you want to build a livable environment—especially one far from the Earth—you must take into account all aspects of human life, which includes food production, energy and oxygen, the construction of shelters, the provision of instruments and the enabling of communications. Teams from different universities were developing oxygen and electricity production processes, a mechanism to cultivate algae, a system to use urine to fertilize plants, sports facilities, a laboratory, a system of communication and control, a robot for digging paths in the ice and more.

Teams were either mono-disciplinary or cross-disciplinary but they worked in the context of a *multidisciplinary project* where different contributions were required to be integrated. Expertise involved ranged from a different kind of engineering, aerospace, biologic, electronic, robotics to computer science, geology, design and architecture. Success in such a multidisciplinary, inter-university and inter-cultural project can only be achieved when each team takes ownership of the overall outcome and takes maximum responsibility for its module, the interfaces of its module and coordination with all teams.

As such, the organizational structure was kept flat and direct communication between teams was preferred over a strict hierarchical organization. With such purpose, regular meetings, and status updates are arranged, as well as a networking communication system based on the app *Slack*. Another useful tool was the *group discussion*, consisting of a virtual meeting on *Webex* involving all the teams. It proved to be pivotal in order to have an overview of the projects, which focused on specific topics, like the exhibition hall setting, or more generic on the whole development, without the need for all moving to a commonplace. Events were usually meant to gather together all the teams and facilitate the information exchange and coordination of different projects.

Our students participated in three events within the network, plus a series of digital contacts with coaches and external experts provided by the *Igluna* network itself. The first meeting took place at ETH Zurich on 12–14 September 2018, and represented the kick-off of the project. The second, on 16–18 January 2019, in CERN, Geneva, aimed instead to assure a conversation among the teams in correspondence of the freezing of all the designs before the start of the manufacturing phase. Furthermore, it was compelled to the delivery of the *Critical Design Review's report*, which allowed us to obtain useful feedback for the next stages. The third event was the *Field Campaign* at Matterhorn Glacier in Zermatt, from 17 June to 3 July 2019 to test parts of *analogues*[11] and all the prototypes that belong to various experiments.

[11]Designing for Space the term *analogue* is used as a synonym for *demonstrator* of Space environments reproduced on Earth in territories with similar characteristics, like the desert to experiment the Mars surface, and the glacier to test the Moon conditions.

4.6.3 Igluna Project's Approach and Value

Academy's students gain access to three main benefits. First of all, it is an important opportunity within a professional perspective: the possibility to experience and report on a curriculum, the participation in a project for Space is intended to boost the future career in this field. Although, even outside the space exploration pathway, *Igluna* project represents an important *training* in terms of *project management*, being compelled to the accommodation in a network of many different teams and contrasting aims.

Moreover, the network itself is an extremely valuable asset, which gathers together not only students and professors from important foreign universities, but also experts of different kinds and institutions like the Swiss Space Center and the European Space Agency (ESA) itself. In this context, they need to behave professionally, learning how to work in a complex system, even exiting their comfort zone at need, but respecting each other skillset.

From a more technical point of view, the approach required is generally a *pragmatic mindset*, aiming first of all to successfully implement the technology developed in the context of the Matterhorn Glacier.

On the other side, the projects exhibited in *Vernissage Art Gallery* were inherently more *conceptual* and focused on the actual lunar environment. Although, they had to keep a scientific approach characterizing the overall network. Particularly, our project, being a link between the actually built *habitat* and its *vision* on the Moon, tried to picture a lunar habitat and mission development that could appeal imagination, but restraining creativity to plausible technologies and operations, based on solid case studies.

4.6.4 Moony. A Lunar Base in Lava-Tubes

It was in this context described above that the project *Moony*[12] was developed: a collaboration between the most prestigious European universities, coordinated by the Swiss Space Center Esa_Lab@, which has as its objective the creation of a lunar habitat demonstrator, complete in all its parts.

The *Moony* project (Dominoni 2019), within the programme *Igluna*, sets itself as a target last one to report the prototypes of all the other European teams involved in the project, in a lunar context. These projects in fact were prototyped first to demonstrate its effectiveness in an environment extreme, such as the Matterhorn Glacier in Zermatt, which has conditions like climate and environmental, which

---

[12]Moony—a whole modular lunar base located under the lunar surface, in the natural tunnels of volcanic origin called *lava tubes*—was designed by Irene Zaccara and Emilia Rosselli Del Turco under the supervision of professors Annalisa Dominoni and Benedetto Quaquaro, experts in Architecture and Design for Space, at the Design School of the Politecnico di Milano.

Fig. 4.41 The lunar base *Moony* built inside the lava tubes with inflatable and 3D technologies using regolith for the protective external structure. Credits by Politecnico di Milano

impact the feasibility, similar to those of a lunar volcanic cave. The aim was, therefore, to create a habitat which serves as a container for projects of other teams. It was, therefore, a question of creating which flexibility can allow incorporation of several projects, still in the phase of development, which will end in June 2019 with the *Field Campaign*.

There are several issues addressed, such as how to build and place a housing module safely in an extreme context responding to requirements that can encourage the adaptation and comfortable stay of human beings.

Our work started with a broad research about the Moon contest, the existing case studies of habitat in extreme conditions, and the related technologies. This led us to the construction of a database of potential technologies and operations that could be used. Then, we applied a decision-making process to make all the *bricks* fit within a scenario, starting from the logistics of bringing the components on the Moon, passing through the construction of the habitat, its protection in emergency situation and finally the actual life inside.

In the first phase, our trade-off verged on the need to *reduce volume and weight*, while enhancing *habitability* and *protection*. The lunar habitat module is designed to connect to other similar space modules in order to create a lunar village and consists of a *semi-spherical inflatable chamber* covered by a light structure that protects it from any collapses that might occur, being the lunar base built in a tunnel underground called *lava-tube*.

This structure it is foreseen to be built In Situ Resource Utilization (ISRU) with 3D printing, using native material, i.e. *lunar dust*—based on a technology developed by the Italian company D-Shape, which had already collaborated with ESA for the *3D Priting Building Blocks for Lunar Habitation* designed by the architect and designer Norman Foster—to minimize the material launchedfrom the Earth.

Fig. 4.42 The picture describes the installation process to build a lunar base in lava tubes and the external structure's installation path to reduce critical elements and guarantee security. Credits by Politecnico di Milano

The use of an *organic design* assures structural performance within lightness and speed of installation (Fig. 4.41).

Moreover, the parametric design can be adjusted based on the currently unknown requirements.

Secondly, we focused on the need for *automation of the construction* and how to simplify at best the process to avoid unexpected behaviours during the assembly. We designed the process step by step and added expedients to overcome potential risks, for example, a track to guide the inflatable process.

Finally, we studied the psychology related to isolation, confinement, and cohabitation in order to be able to suggest solutions that will able to ease the astronauts' pressure about them. This involves a core role of vegetation in the lunar habitat, and a reconfigurable structure, which lets the astronauts be in control of the wide space division without constraining the individual to continuous interactions with the rest of the crew.

Inside the *Moony* habitat, the working, personal and social activities of the astronauts integrate with each other thanks to a *flexible configuration* being able to adapt to its inhabitants according to the various needs. The walls, which are also inflatable, can be opened and closed to transform the environments—by enlarging or reducing them according to the functions to be accommodated—reducing the perception of constraint and confinement in the absence of natural stimuli.

Just to ensure a relationship of well-being between man and nature has been inserted an evolved greenhouse at the centre of the housing module.

Personal spaces for the crew are assured in the so-called *crew quarters*, where each member of the team can enjoy their own environment, although very small in size. The common area, which integrates with the greenhouse, provides small individual corners, where you can live your free time with greater freedom, while maintaining a kind of privacy necessary to escape from the condition of continuous coexistence. During social moments, the same common area becomes an informal place to meet and organize activities of relaxation and entertainment.

4.6.4.1 Lava-Tube: The Ideal Environment

Starting from the assumptions given by *Igluna* project, it was established that the first lunar module was placed inside a tunnel just below the lunar surface. Since sunlight never reaches caves of volcanic caves, if not exactly under the skylight, the temperature remains constant at −20 °C throughout the year and the environment remains permanently dark. The area remains protected from cosmic radiation solar, thanks to a thick layer of about 10 m which separates the tunnel from the outer surface. Naturally placing the habitat underground immediately resolves the problem of free fall of meteorites on the ground, which otherwise would continually endanger the strength of the structure.

Despite all these obvious advantages, the positioning of the lunar module inside the tunnels also carries a particular risk that should be carefully considered: the possible collapse of the ceiling. The discovery of volcanic caves in fact took place with the detection of strange depressions linear that marked the ground for miles and kilometres. These are the tracks left by ancient tunnels all collapsed. Having evaluated this list of strengths and elements of criticality, they are deduced indications well-defined with respect to design of the external structure of the lunar module. First of all, this structure must of course be pressurized due to the lack of atmosphere outside. It must be able to thermally insulate the interior environment, in so that a temperature can be maintained internally stable around 18 °C. The insulation, in this way, will need less protection compared to what is necessary for a lunar habitat located on the external surface that should instead withstand temperature changes of about 250 °C. In addition, the external structure will no longer have the task of protecting the internal environment from radiation, since already the rock layer of the roof, 10 m thick, will perform excellently this task.

4.6.4.2 The Habitation Module

During the design of the habitation module, many factors must be taken into account, as dimensions, thermal insulation and above all the means of transport. An accurate amount of living space per person, the *minimum spaces*, is fundamental for the success of the mission. In the early space missions, NASA had designed environments with only 0.57 m^3 per person, but over the years have been carried out many psychological studies on the negative effects that result from prolonged stay in confined spaces so small. As a result of new research on living spaces for long periods and in confined environments, the minimum volume is of 120 m^3 per person (exactly the same volume considered for the International Space Station).

So, the total volume should be around 360 m^3, calculating the continuous presence of three members of the crew inside the habitation module.

How to carry a module of such size, without greatly affecting the travel costs? Please note that at each pound (0.45 kg) launched on the Moon, corresponds $10,000 worth of fuel. As regarding all the study cases considered, the possible types of module viable housing are

- a rigid module;
- a module built with materials available in situ;
- an inflatable module.

The *Moony* concept emerged from a decision-making process based on the identification of alternatives more plausible or, at least, more convincing. Since we could place the habitat in a *lava-tube*, we could minimize the need for a protective shell. We adopted the solution of inflatable structures and we tried to extend its use as much as possible to all the interior environments, thanks to its practicality in logistics. We looked for a technology for the exploitation of *in-situ resources* for regarding the coverage. We have decided to maintain a single volume to have all the services necessary from the first mission, and to reduce the complexity and risk of failure of a connection between the parts. However, we have included a scalability system. Regarding the interiors, we have assumed to have structures in our habitat supporting all the crew activities living and working inside the lunar base, as crew areas for *entertainment*, resting and personal hygiene, food preparation and consumption in convivial mood, or physical exercise and recreation. In addition, we have assigned a fundamental importance at the *greenhouse*, which becomes the central core of the habitat and characterizes the common life, and we have maintained a large and open *space for sociality*, also thanks to the use of structures deployable. It is important to point out that given the complexity of the project our contribution has focused on the design of the lunar base and mainly on the structure, leaving room for subsequent projects regarding the whole design of the interior and furniture.

4.6.4.3 External Structure's Installation Path

Landing

The lander slowly descends to the ground activating retro rockets. Download the two rovers, the habitation module closed, formed by the two airlocks and the tunnel deployable evacuation.

Descent into the Lava-Tube

Construction robot fixes deployable ladder on a safe crater area and lowers inside of the lava-tube (Fig. 4.42). A mobile system allows to explore the uneven floor until finding a site suitable for habitat, based on a list of requirements, such as a relatively flat, some distance from the skylight, proximity to strategic resources, etc.

Situ Setting

Construction robot analyses 3D morphology to calculate the necessary adjustment for the print and create a negative form CAD to fill the irregularity. Reaches the surface to load enough regolith and start the print. The ultimate goal is to create a flat surface for the habitation module, which leads automatic inflation, an exit area for the placement of the lift and a path free of obstacles linking the two elements.

Descent of the Module

Once construction robots have analyzed the uneven soil and printed 3D a smooth pavement for access and a base, the housing module is lowered into the whole cave, transported and positioned at the centre of the base by the two robots.

Inflation of the Module

Thanks to 3D printed tracks during the preparation, the robot is able to precisely position the habitation module in the area. Following the guides, the inflation takes place in a controlled way. The habitation module assumes naturally a precise and stable location on the site.

Cover and Way Out

After storage of the necessary quantity of ice and lunar regolith, the protection structure is printed with both the autochthonous materials. The habitat always maintains two exits, in order to always ensure a rapid escape from each point. Even when they are connected to multiple modules, two shall be the secured outputs, which will have to increase in number when modules will increase. A LED chain connected to a generator autonomous highlights the exits and the route. The presence of a rigid structure ensures protection until complete evacuation from the habitat. However, a monitoring system constantly measures any seismic signs of potential collapse.

Elevator

The elevator is initially hanging from the crater and extended to the ground below. Next is built with a 3D printed structure only in regolith, so that contact with the crater comes removed and the elevator is independent, not affected by any collapse or collapse of the ceiling of the tunnel. A flexible cylinder, consisting of a sequence of rigid sections connected through a fabric, is prepared to follow the exit path. A 3D printed coverage is created using the cylindrical structure as support.

Walls

Then all the walls will swell partitions, which are integrated into the inflatable but have separate air chambers. The hermetic doors will be brought from Earth and mounted subsequently. The walls are designed with special recesses designed to contain all materials technicians, boxes and personal items of each gender. This will not be necessary to add shelves or furniture.

Furniture

At the arrival of the astronauts are positioned the latest equipment and furniture. Based on principles of collapsibility, lightness and multifunctionality, the entire habitat is completed with objects supporting the daily life created to maintaining an emotional connection with the Earth environment and personal memories.

The furniture is designed to be mounted in the simplest way possible directly when the astronauts arrive in the module. The concept consists of transporting of the Moon light aluminium frames, preformed linings of all furniture and polymer resins expanding. These resins once mixed together, they expand and reach up to 6 times the initial volume. They allow for soft cushions, but according to needs or preferences of the astronaut, you can add small amount of lunar sand to increase stiffness of the object.

4.6.4.4 The Securing the Lunar Base

In addition to the inflatable module, it is, therefore, necessary to address the problem of protection against rock collapses from the *lava-tube* tunnel vault. The hyperbaric structure is very resistant, but it still needs another shielding. To tackle this problem, we have been inspired by many case studies—among which the Norman Foster Lunar Habitation Study Programme for the European Space Agency (ESA) *3D Priting Building Blocks for Lunar Habitation* (2012), followed by the NASA Mars 3D Printed Habitat. In these projects, the habitation modules are often enclosed in envelopes more than one metre, so as to protect the whole module from the extreme environmental conditions that we are forced to face on the Moon surface. Having placed our habitation module in the volcanic caves do not need to be built such shielding, but a structure is sufficient resistant and light. Transport from the Earth a rigid structure that can cover the whole habitation module is an extremely expensive option (0.45 kg = $10,000 in fuel). The best method is to build this cover exploiting the resources that can be found above and below the lunar surface.

The structure does not completely cover the dome, but wraps it in a branched way, in order to speed up the production process. The proposed design is 3D printed, with the sintering system. To process the project the comparison with the company D-Shape was crucial.

4.6.5 The Field Campaign and the Next Edition Igluna 2020

I would like to conclude this case study by including two images that we took during the *Field Campaign* organized with events and exhibitions inside the Matterhorn Glacier in Zermatt, Switzerland.

Fig. 4.43 Benedetto Quaquaro, in the middle between Irene Zaccara (left) and Emilia Rosselli Del Turco (right), and I, inside the Matterhorn Glacier in Zermatt, Switzerland, during the *Field Campaign* for the 1st edition of *Igluna* 2019. Credits by the author

The events were split into two areas, one in the glacier, with the actual demonstrator (Fig. 4.43) and a cinema lounge, available for the projection of videos, and the second one in the town, at *Vernissage Art Gallery*, with an exhibition collecting visuals and explanations from some of the projects (Fig. 4.44), not involved in any experimentation.

It was a great joy for us to be able to participate and support our students through a project that allowed them to open up to the outside world, to present the *Moony* project to the Space scientific community, including astronauts and exponents of the European Space Agency (ESA), which have enriched with their presence an event already full of potential and unprecedented experiences.

This year we participated in the second edition of *Igluna* 2020 that has just ended, unfortunately for students in virtual mode, continuing to deepen the issues related to habitability on the Moon: after making the mock-up *Moony* for the second edition we focused on the preparation, consumption and storage of food in the lunar base, as an essential element for the survival of the crew, but without neglecting aspects of conviviality and tradition to keep strong the connection with the Earth and our memory using researches conducted for the autonomous growth of plants and vegetables in reduced gravity.

We are already considering for the third edition of *Igluna* 2021 to explore the theme of light, given the promising results of *Space4InspirAction (S4I)* this year. The quality of light is very important for the health and well-being of astronauts, who live deprived of natural sensory stimuli with the consequent alteration of circadian rhythms, fundamental for the psycho-physiological balance. Going further would be interesting to also recreate the atmosphere of natural light in a lunar base with reflections and shadows to enrich the experience and remember home.

Fig. 4.44 Benedetto Quaquaro and I (left) and Irene Zaccara and Emilia Rosselli Del Turco (right), at *Vernissage Art Gallery* in Zermatt, Switzerland, during the *Field Campaign* for the 1st edition of *Igluna* 2019. Credits by the author

The conquest of Space is now a reality that is taking shape in an increasingly concrete and fascinating for the younger generations, and projects such as *Fashion in Orbit*, *Space Fashion Design*, *Space4InspirAction* and *Igluna* help to direct young people and the whole society towards a shared and livable interplanetary future and in which the Design will always have a fundamental role, if not indispensable.

References

Clearwater Y (1990) Privacy. In: Space habitability. Avignon working days report. Exploratory Studies Program for the Future European Manned Space Infrastructure (EMSI). Proceedings of American Institute of Aeronautics and Astronautics (AIAA)

De Monchaux N (2011) Spacesuit: fashioning Apollo. MIT Press, Cambridge, Massachusetts, USA

Dominoni A (1998). IVA clothing support system. In: Proceedings of space exploration and resources exploitation, pp 0015.1–0015.4, ESA ExploSpace workshop, 20–22 October 1998, Cagliari, Sardinia, European Space Agency (ESA), Noordwijk, WPP-151

Dominoni A (2001a) Vest. Development of an integrated clothing system for the International Space Station (ISS). In: Proceedings of La Scienza e la Tecnologia sulla Stazione Spaziale Internazionale (ISS), 16–18 May 2001. Centro Congressi Lingotto, Turin, ASI Italian Space Agency, Rome

Dominoni A (2001b) VEST. Development of an integrated clothing system for the ISS. In: Monti R, Bonifazi C, (ed) MSSU, Microgravity and space station utilization, vol 2, no 2-3-4. ASI Italian Space Agency. Liguori Editore, Napoli

Dominoni A (2002) Industrial design for space. Silvana Editoriale, Milan

Dominoni A (2003a) Conditions of microgravity and the body's second skin. In: Fortunati L, Katz JE, Riccini R (ed) Mediating the human body. Technology, communication, and fashion. L. Lawrence Erlbaum Associates, Mahwah, New Jersey, USA, pp 201–208

Dominoni A (2003b) Aesthetics in microgravity. In: Katz JE (ed) Machines that become us. The social context of personal communication technology. Transaction Publishers, New Brunswick, New Jersey, USA, pp 277–284

Dominoni A (2005) VEST. Clothing support system on-orbit validation. In: SAE technical paper series, ISSN 0184-7191. Proceedings of ICES 35th international conference on environmental systems, Rome, 11–14 July 2005, SAE Society of Automotive Engineers, Warrendale, Pennsylvania, USA

Dominoni A (2009) Creative network systems for innovation. In: Proceedings of INTED 2009, international technology, education and development conference, Valencia, 09–11 March 2009, Association of Technology, Education and Development (IATED) , pp 002572–002580

Dominoni A (2008) Creative design practices. The creative process in product design andenvironment for educational. Esercizi creativi di design / Il processo creativo del design delprodotto ambientale per la didattica. Maggioli Editore, Politecnica Editorial Series, Milan

Dominoni A (2010) Cosmic space as a key for creative design. How industrial design in cosmicspace is useful to exemplify an innovative methodology for design process in education. In:Proceedings of INTED 2011, international technology, education and development conference,Valencia, 08–10 March 2010, Association of Technology, Education and Development(IATED), pp 002989–002994

Dominoni A (2015) For designers with their head beyond the clouds. Politecnica, Architettura, Ingegneria, Scienze, Maggioli Editore, Milan

Dominoni A (2019) Nuovi Scenari Abitativi per la Stazione Cis-Lunare Gateway e per le Colonie Lunari. In: Kalin S, Raimondo C, Massicci M (ed) L'Avventura dell'Uomo nello Spazio. Dalla Luna a Marte. 2. 2024. Ritorno sulla Luna, vol 2, "Le Scienze", GEDI Gruppo Editoriale, Puntoweb, Roma, pp 088–097

Dominoni A, Quaquaro B (2017a) Future fashion. Space research inspires innovation in fashion. ROOM. Space J 10, Winter 2017:102–107. The Aerospace International Research Center (AIRC). Publishers Press, Vienna

Dominoni A, Quaquaro B (2017b) Dressing up for space. ROOM. Space J 13, Fall 2017: 106–109. The Aerospace International Research Center (AIRC). Publishers Press, Vienna

Dominoni A, Quaquaro B (2020) Distance learning during a Pandemia. How to maintain as active learning approach: the case study of Space4InspirAction course. In: Proceedings of Edulearn20. Connecting technology with education. 12th annual international conference on education and new learning technologies (INTED), 6–7 July 2020 Palma, Spain. Organizator an Publisher: IATED

Dominoni A, Quaquaro B, Pappalardo R (2018) Space design learning. An innovative approach of space education through design. In: Proceedings of IAC 69th international astronautical congress, Bremen, 01–05 October 2018

Jenkins DR (2012) Dressing for altitude. U.S. aviation pressure suits-wiley post to space shuttle. National Aeronautics and Space Administration (NASA), Washington, D.C.

Gazenko OG, Shumakov VI, Kakurin LI, Katcov VE, Chestukhin VV, Nikolayenko EM, Gvozdev SV, Rumyantsev VV, Vasilyev VK (1982). Effects of various countermeasures against the adverse effects of weightlessness on central circulation in the healthy man. Aviat Space Environ Med 53(6):523–530

Green W, Jordan PW (2019) Human factors in product design: current practice and future trends. Taylor & Francis Ltd

Griffin BN (1978) Design guide: the influence of Zero-G and acceleration on the human factors of spacecraft design. NASA-STD-3000, vol 1/rev B

Moore D, Bie P, Oser H (1996) Biological and medical research in space. An overview of life sciences research in microgravity. Springer, Berlin, Heidelberg

Topham S (2003) Where's my space age? The rise and fall of futuristic design. Prestel Publishing, New York

Trabona C (2009) Dal piede di Armstrong l'idea per i miei Moon Boot. In: Corriere del Veneto,20 July 2009

Wichman HA, Donaldson SI (1996) Remote ergonomic research in space: spacelab findings and a proposal. Aviat Space Environ Med 67(2). Alexandria, VA